Light Burdens, Heavy Blessings

Margaret R. Brennan, IHM

Light Burdens, Heavy Blessings

Challenges of Church and Culture in the Post Vatican II Era

Essays in Honor of Margaret R. Brennan, IHM

Edited by
Mary Heather MacKinnon, SSND,
Moni McIntyre, Mary Ellen Sheehan, IHM

Heavy Blessings, Light Burdens: Challenges of Church and Culture in the Post Vatican II Era

Mary Heather MacKinnon, SSND, Moni McIntyre, and Mary Ellen Sheehan, IHM, editors

Franciscan Press
Quincy University
1800 College Avenue
Quincy, IL 62301
PH 217.228.5670
FAX 217.228.5672
http://www.quincy.edu/fpress

Book design, typesetting, and cover design by Laurel Fitch, Chicago, IL.

Printed in the United States of America
First Printing: August 2000
1 2 3 4 5 6 7 8 9 0

Library of Congress Cataloguing-in-Publication Data

Heavy blessings, light burdens : challenges of church and culture in the post Vatican II era ; essays in honor of Margaret R. Brennan / edited by Mary Heather MacKinnon, Moni McIntyre, Mary Ellen Sheehan.

p. cm.

Includes bibliographical references.

ISBN 0819909904

1. Spirituality--Catholic Church. 2. Women in the Catholic Church. 3. Ecology--Religous aspects--Catholic Church. I. Brennan, Margaret R., 1924– II. MacKinnon, Mary Heather, 1945– III. McIntyre, Moni, 1948– IV. Sheehan, Mary Ellen.

BX2350.65 H437 2000
282'.73'09045--dc21 99-056618

Contents

Foreword

by Mary McDevitt, IHM

The church is the classroom in which Margaret Brennan has been a teacher for most of her life. This is the context in which she has inspired, led and encouraged generations of Catholics in North America and elsewhere. It is within the church that she has been both student and disciple. It is here that she has been a seeker and teacher of wisdom, a prophetic voice of truth.

During her life as a member of the Sisters, Servants of the Immaculate Heart of Mary (IHM), she served as novice mistress and then General Superior of her Congregation and President of the Leadership Conference of Women Religious. Subsequently, she spent more than twenty years teaching theology at Regis College in Toronto and at schools of theology across the continent. The lines of her own life parallel some of the most dynamic and difficult years of the church in North America.

Margaret was born in 1924 into the large and energetic family of Henry J. Brennan and Ann Markey of Detroit, Michigan. The family was not overly pious, but the seven children were raised in the deep, rich culture of American Catholicism. Henry Brennan reminded his children that they had been baptized Irish, Roman Catholics, and Democrats! The entire family would attend Sunday Mass and return home to enjoy a large breakfast with many guests. The Brennan home was a place of warm welcome for countless priests and members of religious communities. Margaret's mother was known for her ready hospitality, and her father was respected as the man who had constructed some of the most important church buildings in the Detroit area, including the Archdiocesan seminary in Plymouth, Marygrove College in Detroit, and the Motherhouse of the Immaculate Heart of Mary Sisters in Monroe.

In 1945, at the end of her undergraduate studies at Marygrove College, Margaret decided to enter the Immaculate Heart of Mary Sisters of Monroe, Michigan. It was within this Congregation, committed to the ministry of education, that her finest gifts would take shape and flourish. When Margaret Brennan graduated with a Ph.D. in Theology from St. Mary's College, Notre Dame, Indiana in 1953, she was one of the early pioneers who had entered this new territory for women religious in the Unites States.

As novice mistress for more than ten years, she worked with others in the relatively new Sister Formation program, pioneered by IHM Sister Mary Emil Penet. This was an initiative which sought to provide women religious with the spiritual and intellectual resources for their apostolic vocation within the church. A generation later, the results of these efforts were obvious for all to see, viz., many women religious had become thoughtful, articulate, and highly educated leaders in the American church.

Those were the years when there were as many as eighty novices for the young Sister Benedicta Brennan to instruct. She did so with zeal, conviction, and considerable creativity. She recalls those days as among the happiest of her life, and she relishes the creativity which the entire novitiate would bring to feast day celebrations and dramatic presentations.

Sister Benedicta learned many important lessons in the process of educating young novices. She wondered about the amount of time she had to spend helping them to understand "the rule," since its language clearly did not fit the theology of Vatican II. She asked herself what this could mean. If the point of "the rule" was not more obvious to those who would be living it, then perhaps the language of "the rule" had to be revised in order to make it more understandable. Similar questions about particular teachings of the Roman Catholic Church were being asked by thoughtful persons around the globe.

Soon the winds of change from Vatican II were blowing everywhere in the once comfortable life of the church. Sister Benedicta immersed herself in the study of the emerging conciliar documents and was particularly interested in the new definitions of the church and the role of religious in it. A shift took place within herself as she pondered the Council's insight that the charism of religious life was not so much a gift from the church as a gift from the Spirit for the church. As the world of women religious began to feel the impact of Vatican II, Sister Benedicta was elected General Superior. The very traditional election was conducted in silence in a General Chapter. There were no speeches or position papers. One day in 1966 she was Sister Benedicta, the novice mistress, and the next day she was Mother Benedicta, the General Superior.

Following Vatican II, she became known as Sister Margaret Brennan and soon became widely known for her courageous and imaginative leadership. She called both women and men religious to a renewed commitment to justice, prayer, spiritual integration and theological study. In Detroit she became involved in the renewal of the Archdiocese, spearheaded by John Cardinal Dearden. They were heady and exciting days which saw the formation of parish councils and an increased role for the laity.

The young Bishop Thomas Gumbleton had been appointed auxiliary bishop of Detroit at the same time as Margaret had been elected Superior General. In Gumbleton's estimation, Margaret played a vital role in the transformation of the Church of Detroit into a Vatican II church. "It was a time of crisis in the life of the church," he said. "Immense changes were taking place. Margaret brought enormous energy, great intelligence, and a charismatic presence. We owe her a great debt of gratitude."

With Redemptorist theologian, Bernard Häring, she helped inaugurate the nation-wide House of Prayer movement. Just before Thomas Merton left for his final journey to Asia, she met with him at a monastery in the redwoods of California to discuss what a house of prayer might mean for apostolic religious. Margaret's vision and ingenuity helped to transform an old barn on the IHM motherhouse property into Visitation, an exquisite place of prayer with an outstanding staff of spiritual directors for visitors from all over the world. The Monroe model has since inspired countless religious congregations as they set about to design their own houses of prayer.

From its beginning, Margaret was part of the world-wide effort of the Jesuits to revitalize the Spiritual Exercises and to make them available to women religious and others throughout the church. It was her inspiration to encourage IHM sisters to become spiritual directors. Today IHMs are known far and wide for their acute perception, skill, and compassion in spiritual direction.

The IHMs have long been known for their academic acumen and expertise in countless fields of learning. In a visionary move following Vatican II, Margaret decided to send ten sisters (one for every hundred sisters in the Congregation) for doctoral studies in theology at leading universities in Europe and North America. She saw clearly that if women were to have any influence in shaping the future of the church they had to be able to speak with an educated theological voice. Those first ten voices were joined by the voices of other IHM theologians who followed their lead, prompted by Margaret's inspiration and encouragement.

In assuming the burden of her times, she traveled far and wide. During the 1970s, she went to Vietnam to search out how her Congregation could respond to the needs of the thousands of children who were orphaned by the tragedies of war. Margaret met with Mother Teresa who asked for IHM sisters to come to Calcutta to take the place of the Missionaries of Charity who were responding to the crisis in Bangladesh.

Margaret, however, knew there were other and different crises in her own country. Even as she participated in the intense renewal efforts taking place throughout the United States, Margaret sensed that the process of renewal in the church would not come without its own pain. She has often spoken publicly about the "wedge of truth" which entered into her mind as she paced her office one morning. "If we decide this, then that will follow, and that. . . and that." She quite rightly foresaw that there would be enormous losses as well as gains in the process of renewal. She was wise enough, however, to know that the process could not be stopped; it had to be lived through with faith and hope—for the sake of the church and the world it was called to serve.

By the early 1970s, Margaret Brennan was the obvious and widely-respected choice as President of the Leadership Conference of Women Religious (LCWR) at a crucial time in its history and development. In the name of the religious of the United States, she made many trips to Rome to represent the vision, inspiration, and direction of American Sisters at meetings of The International Union of Superiors General and at consultations with Vatican officials in charge of relations with religious congregations (The Congregation for Religious). More than any other group in the church, women religious took the documents of Vatican II to heart and assumed the enormous task of reevaluating their congregations in the light of the call of the church.

For Margaret Brennan, the meetings in Rome were blessings as she encountered the rich diversity and enormous needs of the international church. They also became burdens when she experienced painful struggles among the members of the church that she loved so much. Every year the representatives of the LCWR made a request to meet with the Pope to ask his blessing on the women religious of the United States. One of their great sorrows was that they were never granted this audience. To this day, Margaret does not know if their request was ever presented to Pope Paul VI. She recalls the day when women religious received yet another refusal—only to read in *L'Osservatore Romano* that a leading American feminist, Betty Friedan, had been granted an interview with the Pope a few days earlier.

Nevertheless, the leadership team of the LCWR continued in their efforts to ensure a faithful and constructive dialogue between the women religious of America and the officials of the church in Rome. Nadine Foley, OP, who would later lead the LCWR, stated that the very demands of the times called forth reserves of strength and conviction from those who held leadership positions: "During a time of great upheaval, Margaret Brennan proved equal to the challenge at a time when directions for the future were not too clear. A leader among leaders at a critical time, she was able to hold a course that honored the authentic elements of religious life and that at the same time supported and furthered needed change in those things that would enable sisters to meet the challenges of gospel mission in the contemporary world."

Back in North America, Margaret Brennan criss-crossed the continent which had become her wider classroom. She lectured on apostolic and contemplative religious life and on the need to incorporate women more fully into church structures and practices. She spoke increasingly about the importance of understanding the significance of culture, its challenge and potential for the life and practice of faith.

Rosemary Ferguson, OP, the Prioress of the Adrian Dominicans, worked closely with Margaret in Michigan and as a colleague in the LCWR. According to Ferguson, Margaret's tireless efforts were marked by greatness, nobility, and fidelity: "Margaret was a *translator.* She could voice, articulate in a unique way, the mystery of the changes religious, and women in particular, were experiencing at that chaotic time. Margaret spoke with authority and was respected."

After ten years of visionary leadership in the IHM congregation, Margaret Brennan was invited to move to Toronto, Ontario, Canada in

1976. In this cosmopolitan and multicultural city, she deepened and extended her interest in culture. Following a year of renewal in theological studies, she was appointed Professor of Pastoral Theology at the Jesuit sponsored Regis College at the Toronto School of Theology. Students from North America, Australia, Africa, Ireland, the Philippines, and countless other places all over the world, came and still come to Regis to study with Margaret Brennan. Her reputation as a wise and prophetic teacher, a challenging mentor, and an ecumenical ambassador is known around the globe. Currently, she is Professor Emerita of Pastoral Theology at Regis College at The Toronto School of Theology.

Margaret has been recognized many times over for her advisory work among innumerable religious congregations. She has helped them to articulate their charisms for these times, shape their constitutions, and understand the demands of an apostolic spirituality. Margaret's inspiring lectures and interpersonal dialogues regularly merit high praise. In particular, Hans-Peter Kolvenbach, the Superior General of the Jesuits, recognized her distinguished service as a teacher and named her Professor Ordinarius. Margaret's long time affiliation with the Carmelites has been recognized again and again by the major superiors of various Carmelite communities.

In addition to her work with students at Regis College, Margaret remains a leader in her congregation. Her latest responsibility is to coordinate a six year process for theological education for the IHM Congregation. By this endeavor, Margaret has pioneered a massive educational program that reflects the broad vision and possibilities of the post-Vatican II church. It is a process which has inspired and encouraged many other religious communities around the globe to develop their own programs of theological renewal. After reading her explanation of the IHM Theological Education Process, the late Joseph Cardinal Bernadin wrote to commend Margaret saying that what the IHMs were doing was exactly what he hoped for in his Common Ground Initiative.

Now in her mid-seventies, Margaret Brennan has many lines on her face. In her classes she points to these lines with pride. "Each line has a face and a name," she laughs. "I love each wrinkle." Throughout her life, Margaret has always been declared a "born teacher." Such a statement is true, but it does not quite capture the rarity and depth of her gifts. She is a scholar who creates and shapes courses with uncommon depth, breadth, and vision. She truly has the power to inspire others. Her friends, students, and colleagues seek her wisdom, spiritual guidance, and prophetic courage. It is no wonder that so many of them wanted to contribute to this volume. May we have the courage to do what Margaret has always done and encouraged others to do: "to live the questions and to live them graciously."

Introduction

Written with appreciation by colleagues, friends, and students, the essays collected in this book reflect the life-long passions and interests of Margaret R. Brennan, IHM. Each author, in one way or another, reflects on the burdens and blessings of church life today in the midst of massive cultural change. Drawing inspiration from the Christian tradition, the writers espouse a transformational view. They reflect with critical hope on the need for change, both in the theology and practice of the church and in the theory and practice of culture as they interact with each other. The reader will also detect in the essays elements for an integrated view of spirituality, centered on discipleship understood as commitment to realizing anew and more fully the liberating message of Jesus in today's cultural context.

All of the writers are situated within the Roman Catholic tradition. They write mostly from the standpoint of white North American cultural experience, though many of them have also had significant experience in other cultures. These perspectives suggest certain limits to the exploration of church and culture in these essays as well as certain strengths. It is hoped that this book, while offered primarily in honor of the life and work of Margaret R. Brennan, IHM, will also stimulate discussion and dialogue among a wide variety of readers who share its commitment to ongoing critical reflection on the intersection of church and culture for the sake of transformational action in today's world.

In the first essay, Roger Haight sets forth a way of thinking about the theme of burdens and blessings in terms of five challenges that exist in Roman Catholic church life today. Undergirding these challenges is the character of discipleship, described convincingly by Joan Chittister in the second essay, not as imitation of the past but as innovation in the present. Both of these authors cite the full inclusion of women into the life of the church as a major transformational issue, a theme explored carefully and creatively by the next several contributors.

Sandra Schneiders writes compellingly of change, both already here and also yet to come, from the contributions of feminist biblical interpretation. Mary Ellen Sheehan describes the changing practice of women engaged in church ministry and the serious challenges it brings to the integrity of the church with respect to coherence between its proclamation and practice. Theresa Koernke reflects on dilemmas and possibilities as women engage in one of the central symbols of church life, the Eucharist. Mary Ann Hinsdale uses the critical category of interruption to interpret the Spirit-driven contemporary women's movement and its implications for transformed theology and church life.

From a psychological point of view, changes in women's consciousness create new horizons for an integrated spirituality. The next three essays take up this important theme. Ronald Barnes engages some contemporary research on male and female development and argues for new forms of dialogue between women and men. Patricia Cooney Hathaway describes the consequences of psychological change in women for the development of a mature life of faith. Joann Wolski Conn reflects on the tradition's theme of darkness and its relevance for women and men who search for spiritual integration in today's changing context.

The next three essays reflect on change and instituted religious life in the church. Ellen Leonard outlines developments in post Vatican II apostolic congregations of women and points to their significance for the wider church. From a historical and canon law point of view, Sharon Holland reflects on challenges today for authority in religious congregations if they are to retain their prophetic character. Constance FitzGerald mines some contemplative traditions on darkness, absence, and presence to identify links between contemplation and transformation and their importance for us today.

The absence and presence of God theme threads through the next two essays which probe creatively the contemporary experience of suffering and death. Gregory Baum explores similarities and differences between Meister Eckart and Dorothy Soelle on prophetic mysticism and gathers insight for interpreting our contemporary experience of God. Moni McIntyre reflects on death from the perspective of contemporary ecological thought and claims its significance as the wider horizon for a Christian understanding of death.

Reflection on culture and ministry and their challenges for spiritual integration permeates the next three essays. Carl Starkloff offers valuable perspectives from history and from his contemporary experience of ministry among North American indigenous peoples about the challenges that inculturation raises for current ecclesiastical practice. Susan Rakoczy describes her African cross-cultural experience in the ministry of spiritual direction and draws out important wisdom for the global church. Annice Callahan shares some practical wisdom on ways to integrate the practice of ministry into a wider understanding of spirituality.

The need to embrace a widened and more integrated spirituality continues in the last four essays, which explore the global economic and ecological challenges to Christian discipleship. Mary Jo Leddy analyzes theories of the relationship of Christian theology and culture and gains insight from the vision and practice of the Czech writer and political leader, Vaclav Havel, for concrete and hope-filled action. Lee Cormie scrutinizes keenly the extent and roots of massive global economic and social disorder and also identifies seeds of hope for justice in a discipleship of solidarity. Amata Miller develops a widened understanding of the Christian tradition on justice by linking the ecological concern of care for the planet with the liberation concern of care for the poor. Thomas Berry reflects poignantly and with conviction on the severe consequences for Christian theology if it fails to pick up the challenge of contemporary scientific knowledge on the structure and functioning of the universe.

The last words in this book come from Margaret Brennan. In her reflection on the burdens and blessing of church life today, she acknowledges how the authors of these essays have written on issues that have touched her life personally and that continue to challenge her today. For the benefit of our readers, we conclude with a bibliography of Margaret Brennan's works, compiled by Elaine Biollo.

The editors wish to thank each of the contributors for their enthusiastic response to this initiative and for their patience in the process of gathering and editing the essays. We wish to thank as well Juliana Casey, IHM and Mary Jo Leddy who labored creatively with us in the months of conceiving this project. Finally, we wish to express our gratitude to the publishers at Sheed and Ward, and especially to Robert Heyer and Andy Apathy for their care in shepherding this project in its middle stages, and to Laurel Fitch for her exceptional attention to detail and care for us as we entered the final stages of production with Franciscan Press. May burdens be light and blessings be heavy for all who contributed to this book and for all who read it and labor from its vision of committed discipleship.

Mary Heather MacKinnon
Moni McIntyre
Mary Ellen Sheehan,
Editors

The Church *as* Burden *and* Blessing

Roger Haight, SJ

The theme of the church as both a burden and a blessing invites superficiality.[1] It is singularly open to facile contrasts. Should one not list first all the things that are wrong with the church and then, by contrast, point to the things that work? If one probes too deeply into the negativities of the church or takes them too seriously, one's critique will seem unhappy and be taken as mordant. But at the same time too much enthusiasm about the church's blessings will surely appear naive. A spontaneous temptation urges one to align its deficiencies with the church as a human institution which, despite its frailties, is nevertheless divine and thus above any final criticism because it mediates God's grace *ex opere operato*. But the church is not divine substantially, not a divine but a human institution. Ecclesiology has always rejected the notion of the church being hypostatically united with God. Christians, therefore, must find the blessings within the burdens and not alongside them; and these blessings will always cast shadows.

It would be an understatement to say that the Roman Catholic Church in North America is going through a difficult period. The problems began after the Second Vatican Council. They are much deeper than the disgrace that the sexual misconduct of too many of the clergy and the instinctive institutional response to hide or minimize the scandal have brought upon it. In what follows I shall enumerate five basic problems in the Catholic Church in North America today. Multiple reasons might be given for why and how these situations have arisen, but the task here does not include analysis of causes or assignment of responsibility. Generically, all of these problems consist in vari-

1 In this essay the word "church" refers to the Roman Catholic Church except where the context makes it obvious that the referent is the whole church. This restricted use of the word for a part of the whole church, that is, a specific communion, is dictated by the subject matter. Moreover, I limit my scope to the Catholic Church in North America, even though one always implicitly includes the whole Catholic Church when dealing with a regional church because of the centralization of Catholicism.

ous ways in which the Catholic Church has failed to keep in dialogue with the world in which it exists and thus finds itself out of touch with cultural reality. The fundamental promise of Vatican II, especially what was pledged in *Gaudium et Spes*, has not been kept.

What then is the meaning of the terms "burden" and "blessing" as they are understood here? Today, as will be noted below, the church in North America is a voluntary association. One does not have to be a member; no social pressure demands it. One is a member by choice, because one wants to belong. In a voluntary association one identifies with the institution of which one is a member and makes its ideal one's own. Many Catholics in North America today have left the church because the weight of many of the Church's official positions, or those publicly perceived to be so, is too heavy to bear. More and more these people cannot assume the responsibility of membership in this church; they do not want to be associated with what it stands for. These burdens cannot be associated with the will of God; they cannot be considered "sacrifices" which, though difficult, sanctify. These burdens can only be blessings dialectically and paradoxically. One turns them into blessings not by accepting them, but by accepting the church in spite of them and through a dedication to change them. Some of these burdens are blessings only because they measure the hope of the members and the courageous dedication of some official ministers to represent another way. What follows, then, are five points outlining ways in which the church has become a burden to its members, because while both they and the world have changed, the church has not. Only paradoxically may they be considered blessings by those who are dedicated to reinterpretation and reform so that the true character of the church may appear.

An unchanging institution?

Educated people in North America today are aware of what it means to live in history. Historical consciousness involves a recognition that we come from a distant past, that whatever exists today developed historically out of the past. Even ordinary people appreciate the particularity of human realities, the differences between things in different times and places, and the relativity of situations, social and cultural settings, ideas, and values. These differences are thoroughgoing and reach beyond the content of thought to differences in the very ways people of various epochs and cultures think. Thus historical consciousness in this deeper sense implies a sense of the relativity of all things, that the content and meaning of everything is constituted by its relatedness to the things around it. There is nothing or very little that is generally or universally true for everyone, because every situation is individual and particular, and every single thing is seen from the viewpoint of distinctive angles and questions. In the course of its continued existence in time reality itself changes. Thus a sense of constant change, of a lack of stability, and of pluralism is concomitant with historical consciousness.[2]

Vatican II was very influential in bringing this historical consciousness into the church at large by dramatizing pluralism and then demonstrating the possibility of change. It also brought a sense of historicity to bear on theology, for this council opened the church up to dialogue with the modern world, its history, and its cultures. And this historical consciousness, in turn, has considerable significance for how we are to understand the church. For in an historically conscious framework it becomes apparent that the church, in all of its developments through history, took on the social forms and ideas of the age in which it existed. But if this is the case, the idea of a God-given and divinely willed hierarchical constitution of the church does not fit with people's experience. If the structure of the church can be completely explained by historical development, and if as a product of history it is always changing, it will be difficult to make sense of divine claims or divine prerogatives for any particular ecclesial structure. The church in its institutional form at any given time is a product of history and human decision.

This historical consciousness forms a context for thinking in our culture today. Catholic people are aware of Protestant, Orthodox, and evangelical Christian churches and are comfortable with the idea that they are authentic and valid Christian churches. Many of their children or relatives have married into these other churches; many of their friends are members. Catholic people are also aware of other religions and tend to view these religions positively. If they live in or around urban centers, their neighbors may be Hindus, Buddhists, or Muslims. That there are many religions is not a negative fact; it is expected and

[2] For a concise description of historical consciousness see John O'Malley, *Tradition and Transition: Historical Perspectives on Vatican II* (Wilmington: Michael Glazier, 1989), pp. 73–77.

valued, for one particular religion could never unfold the richness of the mystery of God and God's dealings with human beings. Vatican II also stimulated this positive appreciation of other religions with its doctrine of the universality of God's grace. Catholics like everyone else have a much broader view of the world today because the world has become smaller; the revolution in worldwide communications has helped to overcome a narrow sectarian view of reality. The church is decentralized in history; it is no longer considered the exclusive place or even the center of God's work in the world. One whose idea of the cosmos has been expanded by the findings of science over the last thirty years cannot consider the Roman Catholic Church the center of human history.

To the extent that people in North America, Catholics in particular, share in this historical consciousness, they will find the ecclesiastical language the Catholic Church uses to describe itself a burden. It is true that for some the church may have changed too much too fast and left them behind. But this is not the case for the great majority of young people and those actively at home in North American society. People who live in today's world are not pining for yesterday's church. A church is a burden for people who have realized that, although the church has inculturated into other places and periods of history, it refuses to do so in this time and place. It continues to use a-historical and absolute language to describe itself, cherishes uniformity rather than an inculturated presence to people in different ways in different places, looks backwards instead of forwards, is uncompromising in negotiating with other Christian churches, is undemocratic its structures, dictates answers without listening to questions, and controls the periphery from a culturally alien center. What is presented here is not simply a catalogue of faults, but a description of an institutional culture which for many is alienating and intolerable. This is a corporate mindset that appears in different ways to different people: triumphal and arrogant, curiously antiquated and quaint, alien and bizarre. How could such a culture be a blessing for members who are products of North American culture, inquisitive about the particular significance of new situations, used to cultural pluralism, forward looking, tentative, adaptive, conciliatory, and adept at listening, negotiating, and building consensus? Only through a negative experience of contrast. On the basis of such a negative experience one would react against this alienated and alienating ecclesiastical culture in order release the positive values that are concealed by ideas and practices that stem from another culture. Ministry today should help people to assume the responsibility assigned the laity by Vatican II for participation in the reform of the church.

An individualistic institution?

Over the past two centuries we have learned that human existence cannot be defined only on the basis of the individual. This is one area in which the enlightened philosophers of modernity were short sighted. Many academic disciplines, from biology, through linguistics and philosophy, to the social sciences show the degree to which human existence is socially and historically constituted. One cannot take the individual person as the imaginative referent for a definition of the human; one must always see personal responsibility in dynamic tension with the social relationships that together constitute concrete human existence. Individuals are also members, co-responsible partners, in a corporate enterprise of being.

The single most significant contribution of the Second Vatican Council to the understanding of the church consists in defining the church in its relation to the world and human history. This occurred principally in its document on the mission of the church, *Ad Gentes*, and in *Gaudium et Spes* on the church in the modern world. The church is an open institution, that looks out upon the world, and understands itself in service to the world because it mediates something of value for the world. The church is not an enclave within the world isolated from the world and closed in upon itself, but essentially missionary, that is, sent to the world.

But when one looks at our world today what one sees is a degree of suffering that is structured socially in cultural, political, and economic forms and, in terms of sheer numbers, is unprecedented in human history. The phenomenon of refugees around the world is a good index of the newness of our world situation. The numbers of people in the world who have been dislocated from their homelands is shocking. The massive poverty, dehumanization, and social oppression is the concern not only of its victims but also of every thinking human person. But the question for the church is this: what does the church have to say or do about this common situation of humankind? The church claims to "possess" or "contain" God's salvation as this is mediated through Jesus Christ. But what has this salvation or grace to do with the common condition of the human race today, that is, concretely and in its actual history? The church cannot say that it is not involved in social and political issues. The church according to Vatican II is a sign to the nations of God's salvific presence to human beings in their history. The church's response to the social injustice that keeps so many people in bondage today is a key to its credibility before the nations. It is impossible today to understand the church in any credible way without formulating this self-understanding in a way that is responsive to global social injustice, poverty, and dehumanization.[3]

But does this describe the Catholic Church in North America? It is almost impossible to generalize and produce an image of the whole church in North America because local churches and parishes inevitably take on the character of the communities they serve. The church is pluralistic. But among much of its middle class constituency, it has taken on the character of a highly sacramental church with an individualistic notion of salvation in reaction against a secularized society. The term secularization is a tricky one. It is particularly hard to apply it consistently to a society such as that of the United States. For in comparison with other Western societies this is a very pious and religious nation. But it is clear that in our culture and society there is a large separation between the lives of the churches and the running of our secular institutions such as business, industry, and government. This separation is by no means complete; there are many points at which the churches directly intersect with society. But some degree of separation between spirituality and secular life is demonstrated by a comparison between our society and, for example, that of Europe during the Middle Ages or some Islamic nations today. In other words, the secular sphere has been differentiated from the religious and is quite autonomous. Regarding the lives of the churches themselves, it is not that society has become less religious; the rise of fundamentalism and the electronic churches has demonstrated the opposite. Rather religious culture has in many

[3] See the collection of essays by Johannes Baptist Metz, *The Emergent Church* (New York: Crossroad, 1981) which develops the implications of a shift from a privatized to a socially conscious faith for the life of the church.

areas become individualistic and even privatistic. In this respect the church has become inculturated into a typically modern Western industrial society.

The Roman Catholic Church has flourished in the United States, but despite its strong corporate identity, its notion of salvation, too, has become influenced by individualism in this sense: salvation has been largely a purely religious and personally moral affair. It is not that being a Catholic has been altogether divorced off from engagement in the world and the assumption of responsibility for social problems. The church has had a strong social ethic and has been an advocate of its immigrant members. But the link between this social commitment and religious spirituality has been extrinsic and consequential. A spiritual relationship with God is considered possible without a social commitment. The degree to which social concern is self-interested shows its autonomy and independence from a religious attachment to God mediated by Jesus. Religious piety is one thing; social action is quite another. People who, like Dorothy Day, see their relationship with God as constituted by a commitment to the poor are somewhat to the left of the center of Catholic spirituality.

Where, then, does the burden lie and who carries it? Not with the new immigrant churches, the hispanic and other "ethnic" communities who are poor and for whom social struggle and religion are scarcely distinct. It lies rather in the middle-class parishes and is shared by those who are dedicated to social welfare and the cause of social emancipation.[4] For these people church appears irrelevant and other institutions attract their religious energy. The burden is really a lack, a void, an absence, a vacuum. Other institutions appear to be contributing more to the emancipation of the human body, mind, and spirit than is the middle-class Catholic Church and its leadership. Other secular organizations which address ecological, social, racial, economic, and cultural problems substitute for the religious commitment of socially conscious Catholics.[5] One senses that the specific religious exigencies of our time

[4] Once again, it is not that such a social doctrine is lacking, it is rather that it is a secret that rarely enters into the actual everyday preaching of the church. I address this in R. Haight, "Liberation Theology and Middle Class America: A Personal Reflection," *Chicago Studies* 32 (April, 1993), 64–76.

[5] The suggestion is not that the church become another social service orga-

are being avoided and that the inner life of the affluent parish consists of a boring, family-centered, individualistic piety.

A SEXIST INSTITUTION?

Very few people in North America are unaware of the feminist movement, in some respects revolution, that has occurred over the past three decades. Today this movement is gaining momentum in Latin America, Asia, and Africa. Whatever one's judgment may be regarding various aspects of this movement, for it is at this time highly differentiated, one cannot fail to be impressed by the witness of women to their own experiences of various forms of discrimination and even oppression. But listening and attending to this witness forces one to examine the objective situation of women in our society today and in societies and cultures throughout the world. I take it as simply a matter of fact that women throughout the world have been and are consistently the victims of discrimination and oppression. From an egalitarian perspective, that is, a perspective expecting that relationships between all people are governed by social institutions that are just, even-handed, and non-discriminatory, one has to say that throughout the world, and generally throughout our past history, women are and have been the victims of injustice and, in many areas, of violence. This consciousness which is well developed in Western culture can only continue to grow world-wide.

Here again a very serious question is raised for Christianity and specifically for the church. The seriousness of this problem can be seen in the following contrast. On the basis of its message about God and God's will for humankind, the church should be at the forefront in opposing all forms of human oppression. But in this case, it is not and has not been. On the contrary, in the eyes of many the church itself is a

---

nization. The church is a church, and social agencies do what they do. What is lacking in the middle-class churches is a socially conscious spirituality and preaching that engages parishioners in other secular organizations out of explicitly religious motives. The agenda of social issues are the material and subject matter of religious spirituality and preaching. See John A. Coleman, "A Church with a Worldly Vocation," *An American Strategic Theology* (New York: Paulist, 1982), pp. 35–56 for a systematic proposal for understanding the church's engagement in the world.

sexist institution. But even worse, the church seems unable to escape from this position because of its own self-understanding. For it is a matter of record that both Jewish and Christian self-understanding has been mediated to us from the past by a history of patriarchal forms, both social-institutional forms and formal patterns of self-understanding. Thus the very tradition of the church is sexist. The question for the church, then, is quite sharply defined: is the church able to understand itself in non-sexist terms and thus able to be a light to one half the population of the nations and an actor in this liberation movement?[6]

The burden here is, of course, born primarily by women members of the church. It is simply remarkable that women who are feminist remain loyal to the Roman Catholic Church. But this burden is shared by all who remain members of this institution. It can only be a blessing by providing a cause for which one can commit oneself in the name of the values of Jesus' own ministry.

A FIXATED SEXUAL ETHIC?

No one would dispute the contention that the Catholic Church has strong views in the matter of sexual morality. Since sexuality is a rather central concern in human life and is surrounded in every culture by moral norms and a practical ethos, it is not surprising that it is closely aligned with religious values and precepts. Before the Second Vatican Council the Roman Catholic sexual ethic was finely developed, and in North America, Roman Catholic sexual morality was widely and deeply known and internalized. Parents and young people alike knew the rules, tried to live by them, and were conscious of it when they did not.

It would be surprising to hear Catholics making the same claim today after the sexual revolution. One would not be far wrong to imagine that the sexual behavior of Catholics does not vary significantly from that of other North Americans of the same representative economic and cultural classes. But for all the changes in Catholic sexual

6 Among many works in Catholic feminist theology Anne Carr's *Transforming Grace: Christian Tradition and Women's Experience* (San Francisco: Harper & Row, 1988) is systematic, comprehensive, incisive, and practical. See also Catherine Mowry LaCugna, "Catholic Women as Ministers and Theologians," *America* 167 (October 10, 1992), 238–48.

attitudes and behaviors, the teaching of the Roman Catholic Church remains substantially the same.

Where does the burden, the discouraging weight of disappointment, lie in this area? I am not competent to say with the proper nuance that it lies in the sexual ethical or moral teaching itself.[7] Rather it lies in the enormous chasm between the cultures. In a culture in which a fifty percent divorce rate obtains, serial divorce is accepted, censorship has all but broken down completely, contraceptives are a way of life and aggressively promoted as a factor in alleviating the effects of teen-age pregnancy, abortion, and AIDS, and a homosexual lifestyle is becoming more accepted, more and more people have little idea of what the Catholic Church is talking about in sexual matters. The gulf lies not simply with the values and prescriptions, but in a context for understanding. It is as though the teaching were addressed to another people in another age; there is no medium or context for communication. In short, the church appears out of touch with reality; its official teaching has lost contact with its own members. In the terms of an integral ecclesiology of the people of God, the church's self-understanding in this matter has completely broken down because we have lost a language to communicate among ourselves.

And this is where the burden lies for one who remains in the church and is even willing to assume some responsibility for its public ministry. In this case the metaphor of weight shifts to that of pain or hurt. The feeling is analogous to witnessing one whom one loves gradually lose contact with reality; the pain is a kind of pity at the fear of neurosis or sclerosis that distorts the good intention of the beloved and gradually marginalizes and finally isolates him or her from the everyday course of affairs. So too the burden of the member as well as the local minister is one of sadness as the leadership of the church systematically destroys its own credibility.

AN AUTHORITARIAN INSTITUTION?

As was noted at the outset, participation in the Roman Catholic Church has become a voluntary activity. People in urban and suburban

7 For a discussion of some of the aspects of the disarray among American Catholics regarding sexual matters see Lisa Sowle Cahill, "Abortion, Sex and Gender: The Church's Public Voice," *America* 168 (May 22, 1993), 6–11.

situations are no longer under any social constraint to be active members of any church. Thus even for Catholics the Roman Catholic Church has ceased to be a necessary institution for one's salvation, but has become a voluntary association to which one belongs only if one so chooses. Many Catholics have found that other churches more closely fit their religious needs. This phenomenon is usually expressed this way: in Western secular societies, which share in some degree an historical consciousness and the experience of pluralism, the church, and even being a Christian, no longer have social support. Being a member of the church is more and more a matter of personal commitment.[8]

This social phenomenon and the change that it has wrought on the people who make up the church have significant implications for the notion of the authority of the church, both the way it can be understood and the way it can be exercised. The most dramatic example of a shift in this area in the American Roman Catholic Church is the general reaction to its teaching on sexual morality. In the classic case of *Humanae Vitae* this doctrine was simply rejected, or in more technical terms not received, by a majority of the members of the church. But this case is in some measure paradigmatic. What one has today within the Roman Catholic Church is "pick and choose" Catholics, that is, members who adhere to some doctrines but not all. A former ecclesiological doctrine of authority, which said that not to accept all the doctrines taught by authority was to reject the authority itself, has simply broken down. But this raises anew the question of the authority of the church: what does, or better what *can*, the authority of the church mean today in a secular and individualistic culture? Is there such a thing as an authority which at the same time is subject to interpretation, dissent, non-obedience, and simple disregard, all of which seem to be the ways in which many Catholics now respond to church authority?

Here more than in some of the other areas one can see ways in which the blessing outweighs the burden. There is such a thing as an authority that appeals to freedom, one whose archetype is not military and whose exercise does not demand immediate, external, unthinking, and unfree obedience.[9] This specifically religious authority is God's and

8 See Karl Rahner, *The Shape of the Church to Come* (London: SPCK, 1973), pp. 55–60, 93–101.

9 See James Gustafson, "Authority and Consent in the Voluntary Church," *The Church as Moral Decision Maker* (Philadelphia: Pilgrim, 1970), pp. 126–32;

only touches the internally free human spirit by invitation. In a pluralistic situation, when official authority does not succeed in mediating God's values to a particular, culturally formed conscience of the individual Catholic, he or she remains morally free to ignore the prescription. More and more Catholics are internalizing this traditional teaching of the schools, and this is putting new demands on how authority can be exercised in our culture. It also bestows on the Catholic the freedom and blessing of the Christian to remain in the church in a loving dedication to its continual reform.

CONCLUSION

It seems that I have fallen into my own trap. The burden clearly out-weighs the blessings in what appears to be a list of complaints about the Catholic Church in North America. Let me conclude by explaining why this is not so.

The church is an institution. If one takes a concrete view of the church as a human phenomenon of history, and refuses to theologize about the church without keeping an eye on this social reality, then of all the models that Avery Dulles proposes for the church the one that most surely corresponds to its most basic reality is the "church as institution."[10] If the church were not an institution, neither would it exist. But like all institutions that exist in history, especially old ones that are tradition-bound, the church adapts to history very slowly. The problem today is that the speed with which nations and societies are changing today is itself rapidly increasing. This puts a special burden on the church, one that was exaggerated because of the relative lack of doctrinal development between Vatican I and II and the suddenness with which Vatican II released pent-up energies.

It seems to be a simple fact that in trying to catch up with modern culture, and then during the pontificate of John Paul II trying to stop

---

Joseph Komonchak, "Authority and Magisterium," *Vatican Authority and American Catholic Dissent: The Curran Case and Its Consequences*, ed. by William W. May (New York: Crossroad, 1987), pp. 103–14.

10 See Avery Dulles, *Models of the Church* (Garden City, NY: Doubleday, 1974). I take the term institution here generically. Dulles discounts the church as institution as an viable model because he gives it a negative specificity drawn from post Tridentine Catholicism.

the process, the church has managed in many respects to distance itself from the social reality of North America. Is it the same thing to say that people in our culture are distancing themselves from the church? In either case, what has been called a burden for many in the church today consists in a description of the distance between the church and the progressive elements in North American society.

But institutions finally are made up of people. While aspects of an institution may be self-destructive of its own mission at any given time, there are elements which are not. Many of the church's official and unofficial ministers provide a de facto, concrete, informal, and formal symbiosis between the church as an institution and its people. On the one hand, one must be saddened by the massive distance between ecclesiastical culture and the many people it is supposed to mirror, serve, and nurture, between the institutional church and the groups of people whom it has abandoned or who have left it. On the other hand, one must expect some distance between the church and its constituency in a rapidly changing society.

In this context the blessing of the church lies in living the faith and ministering to people from within the church. The blessing that the church is becomes realized on the level of historical existence, in the parish, and in informal and formal basic ecclesial communities. In short, the church is actualized in love of God and love of neighbor. In some respects the blessings mediated by the institutional church must be realized despite the institutional church as it is. But the active life of faith provides a basis for what the church should become, that is, an institution growing out of and being responsive to the faith of a people. The blessing bestowed by the church ultimately comes from God as mediated by Jesus of Nazareth. It can only be appropriated existentially through concrete life and ministry. This life and ministry should always assume some responsibility for changing the church so that it corresponds more closely to the values of God as revealed in the ministry of Jesus. As was said at the outset, the blessings must be found not alongside but within the burdens.

Discipleship: The Questionable Measure *of* Christianity

Joan D. Chittister, OSB

The poet Basho writes: "I do not seek to follow in the footsteps of those of old. I seek the things they sought." It is a lesson in discipleship that we may all need to learn again if the church itself—from papacy to people—is to be counted among the disciples of Christ.

Today, as seldom before, the church, both as institution and as the people of God, is being forced to choose between imitation and discipleship. To confuse one with the other is to skew the whole Christian enterprise. The only really meaningful question for a Christian is how to make Jesus relevant, real today. At one level, that can't be done. The Jesus of history does not live in our time. At another level, if Christianity is to have any meaning at all, making Jesus relevant now must be done. What Jesus came to be, to do, to create must be made real in us today or Christianity does not live at all. But to do that it is necessary to consider, first, what the notion of discipleship really implies and second, what the call to discipleship demands now. The two dimensions are inextricably linked: what discipleship implies is clearly what it demands. The concept and the call are of a piece. If not, discipleship becomes either play-acting or fadism. In the first case we will take discipleship to mean that we must do only what Jesus did—as if we could. In the second case, we will do only what we feel like doing—as if we should—and say we do it because of Jesus.

One of the major questions facing the institutional church today is the place of women in the life of the Christian community. Some people argue that in answering the question of the time, we can only do what Jesus did. Others maintain that we can't minister in this world at all unless we seek what Jesus sought—equality, justice and all-inclusive love. The issue hinges, I think, on the theology of discipleship that energizes the people of God: priests, people and popes together.

The facts speak for themselves. Christian discipleship is a very dangerous thing. It has put every person who ever accepted it at risk. It made every follower whoever took it seriously on alert for rejection, from Martin of Tours to Dorothy Day, from the early church to the present time. It cast every fragile new Christian community in tension with the times in which it grew from then to now.

To early Christian communities it meant to defy Rome, to stretch Judaism, to counter pagan values with Christian ones. It demanded very concrete presence; it took great courage, unending fortitude and clear public posture. It took the rejection of emperor worship, the forswearing of animal sacrifice, the inclusion of gentiles, the supplanting of dependence on the Law with a commitment to love recklessly. And all. And everyone.

The following of Christ was not an excursion into the intellectual, the philosophical, the airy-fairy. It was real and immediate and cosmic. "Come, follow me," became an invitation to defend the poor, an obligation to peacemaking in warring societies, a call to identify with the outcasts of society, an experience of public disapproval and personal exclusion. Discipleship became the step-over point between philosophy and sanctity.

The problem with Christian discipleship is that instead of simply requiring a kind of academic exercise or personal piety—the conventional understanding of most kinds of "discipleship," religious or otherwise—Christian discipleship qualifies as authentic only when it demonstrates publicly what it claims to hold on a personal level. Christian discipleship requires a kind of living that is sure then, eventually, to tumble a person from the banquet tables of prestigious boards, the reviewing stands of presidents, and the parliaments of empires to the most suspect margins of society. Because the disciple of Christ looks for a new world order based on right relationships, on justice and on love—on a Jesus-perspective—in a world that builds social relationships on personal gain, personal life on acquisition and the civic order on charity without social change, discipleship invites tension. To follow Jesus, in other words, is to follow the One who turns the world upside down.

It is a tipsy arrangement at the very least. People with high need for approval, social status, and public respectability need not apply. "Following Jesus" leads always and everywhere to places where a person would not go, to moments of integrity we would so much rather do without. It is not doing what Jesus did; it is seeking what Jesus sought.

The Christian, then, carries a world-view that cries for fulfillment now. Christian discipleship is not preparation for the hereafter. It is the commitment to live now as if the Reign of God had already come—so that it may. To follow Christ is to set about fashioning a world where the standards into which we have been formed may well become the standards we find that we must ultimately forswear. Flag and fatherland, profit and power, chauvinism and sexism done in the name of Christ are not Christian virtues whatever the system that looks to them for legitimacy. To follow Christ is to honor all creation, not simply the country in which we are born; to profit the other, not simply the self; to cross boundaries and recognize all of humanity as valid and valuable. Christian discipleship is not about doing what Jesus did. It is about living in this world the way that Christ lived in His.

It is a very demanding process. Discipleship implies now, just as it did then, a commitment to leave nets and homes, positions and personal security to be now in our own world what Christ was for his—healer and prophet, voice and heart, call and sign of the God whose design for this world is justice and love, who hears the poor Lazarus and ministers to the abandoned Hagar. Christianity at its best confronts a world bent only on its own ends with a picture of what a new world based on the heart of Christ would look like, whatever the cost. The price is a high one. Discipleship cost Martin of Tours his status, Dorothy Day her reputation, and Martin Luther King Jr. his life.

The problem is clear: to claim discipleship, the church must not only preach the gospel, it must model it. And surely it must never obstruct it. Religion that colludes with the dispossession of the poor in the name of patriotism becomes just one more instrument of the state. Religion that blesses oppressive governments in the name of obedience to authority makes itself an oppressor as well. Religion that goes mute in the face of massive militarization practiced in the name of national defense abandons the commitment to the God of Love for the preservation of the civil religion. Religion that condones the pauperization of women in the name of motherhood and denies the ministry of women in the name of God's will flies in the face of the Jesus who overturned tables in the Temple, contended with Pilate in the palace, chastised Peter to put away his sword and, despite the theological teachings of the day, commissioned women to preach His name and raised a socially useless girl-child from the dead.

Clearly, discipleship is not based on civil quietism and private piety. Still less is it based on preaching one thing and institutionalizing another.

On the contrary. Discipleship confounds right reason and good sense with right relationships and good heart. It pits the holy against the human. It pits the heart of Christ against the heartlessness of an eminently sensible world. To be a disciple is to find ourselves in contradiction. We become purveyors of a world where the weak confound the strong. We come to proclaim that humility raises a person up and pride destroys. We begin to seek a world where the rich and the poor change places. We set out to shape a world where the last are made first. We insist on a world where women have the right to do what heretofore has been acceptable only for men because they are humans, because they, too, have been called to discipleship. The Reign of God becomes a foreign land made home. "Come follow me" becomes an anthem of public proclamation and human liberation.

Discipleship, we know from the life of the Christ whom we follow, is not membership in a social club called a church. Discipleship is not an intellectual exercise of assent to a body of doctrine. Discipleship is an attitude of mind, a quality of soul, a way of living that is not political but which has serious political implications, that changes things because it simply cannot ignore things as they are, things that defy the will of God for humanity.

The disciple takes public issue with the values of a world that advantages only those who do not need to be advantaged, takes aim at institutions that call themselves freeing but which keep half the people of the world in bondage, takes umbrage at systems that are more bent on keeping improper people out of them than they are in welcoming all people into them, takes the side of the poor always despite the power of the rich. Discipleship cuts a reckless path through corporation-types like Herod, institution-types like the pharisees, system-types like the money-changers and chauvinist-types like apostles who want to send the women away.

Discipleship stands bare naked in the middle of the world's marketplace and, in the name of Jesus, cries aloud all the cries of the world until someone, somewhere hears and responds to the poorest of the poor, the lowest of the low, the most outcast of the rejected. Anything else, the gospels attest, is certainly mediocre and surely bogus discipleship. Christianity is human community writ large and discipleship is its voice.

It is one thing, then, for an individual to summon the courage it takes to stand alone in the eye of a storm called "the real world." It is another thing entirely to see the church itself be anything less than the living Christ. If the church is to be relevant, if the church is to be a community of disciples, then it must be inclusive. To see a church of Christ deny the poor and the outcast their due, justify the very systems in itself that it despises in society, is to see no church at all. It is at best religion reduced to one more social institution designed to comfort the comfortable but not to challenge the manacles that bind humanity to the cross. In this kind of church, the gospel has been long reduced to the catechism. In this kind of church, prophecy dies and justice whimpers and the truth becomes too dim for the searching to see.

The church stands at the crossroads of discipleship today, choosing between community and club, between the prophetic and the institutional, between the gospel and a law designed to cut half the world out of their spiritual birthright if baptism is really baptism at all. Today, as never before in history, perhaps, the world and therefore the church within it is being stretched to the breaking point by life situations that, if for no other reason than their immensity are shaking the globe to its foundations. New life questions are emerging with startling impact and relentless urgency. The poor are crying for humanity. The outcasts are demanding a seat at the tables of the world. The living dead are wailing resurrection songs. And the greatest challenge of them all, the one that is touching every part of the world, every home, every institution, every world issue, is the woman's question.

The church cannot be a community as long as women remain invisible, rejected and reduced to consumers of the faith instead of being able to function as fully adult Christians in a community of disciples. Women are most of the poor, most of the refugees, most of the uneducated, most of the beaten, and most of the rejected of the world. They are also most of the exalted. On no other class, surely, has so much poetry, so much music, so many flowers, so much adulation, so much tolerance, so much romantic love and so little moral and intellectual, spiritual and human respect been lavished. Now we even read encyclicals of praise and apology about women which then invariably end by reiterating their "special" nature, an empty paean which translates to mean their separateness, their second-classness, their subservience, however nice the words. Where in that is the presence of Jesus to the homeless woman, to the beggar woman, to the abandoned woman, to the ministering woman, to the woman alone, to the woman whose questions, cries and life experience appear no place in the systems of the world and no place in the church as well? Except of course to be defined once more as another kind of human, not quite as competent, not quite as valued, not quite as human, not quite as graced by God as men? Where in that is the discipleship of Jesus? Where in that is living in today's world as Jesus lived in his?

What does the theology of discipleship demand here? What does the theology of discipleship imply here? Are women simply half a disciple of Christ? To be half commissioned, half noticed, and half-valued? Or are women the other half of the church, without which the Christian community will never be whole and its discipleship will never be authentic?

In the light of these situations, there are, consequently, questions in the Christian community today that cannot be massaged by footnotes nor obscured by jargon nor made palatable by the retreat to faith. On the contrary, before these issues, the footnotes falter, the language serves only to heighten the problem, faith itself mocks the question. The discipleship of women is the question that is not going to go away. Indeed, the discipleship of the church in regard to women is the question that will, in the long run, prove the church itself.

In the woman's question the church is facing one of its most serious challenges to discipleship since the emergence of the great and agonizing debates about slavery. Tradition, it seemed, affirmed slavery; major philosophers had accepted slavery; an apostle himself preached about slavery. But, in the light of new evidence, under the scrutiny of new thinking, in the face of the piercing example of Jesus, discipleship demanded something new, demanded human freedom, demanded taking people in who had once, by the Will of God some claimed, been kept out.

The major question facing Christians today, too, is what discipleship means in a church that does not accept women as full followers of Christ. If discipleship is reduced to maleness, what does that do to the rest of the Christian dispensation? If only men can really live discipleship to the fullest, what is the use of a woman aspiring to discipleship at all? What does it mean for the women themselves who are faced with rejection, devaluation and a debatable theology based on the remnants of a bad biology that has theologized women out of theology? What, as well, does the rejection of women at the highest levels of the church mean for the discipleship of men who claim to be enlightened but continue to support the very system that mocks half the human race? What does it mean for the church that claims to be a follower of the Jesus who healed on the Sabbath and cured women with an issue of blood? And finally what does it mean for a society badly in need of a cosmic world view at the dawn of a global age? The answers are discouragingly clear on all counts. Christian discipleship is not simply in danger of being stunted. Discipleship has, in fact, become the enemy. The refusal to admit women to full discipleship—something the church itself teaches is required of us all—has become at least as problematic a matter for the integrity of the church as the open discussion of its possibility could ever be. The fact that those who are admitted to discipleship continue to exclude women from the offices of the church that shape its theology, guide its spiritual life and minister to its people, let alone from ordination itself, brings the very notion of discipleship itself into question. Women are beginning to wonder if discipleship has anything to do with them at all.

And therein lies the contemporary question of discipleship. Some consider faithfulness to the gospel to mean doing what we have always done. Others find faithfulness only in being what we have always been. The distinctions are crucial. The distinctions are also essential to the understanding of discipleship in the modern church.

When "the tradition" becomes synonymous with "the system" and maintaining the system becomes more important than maintaining the spirit of the tradition, discipleship shrivels. It becomes at best "obedience" or "fidelity" to the past but not necessarily a deep-down commitment to the presence of the living Christ confronting the leprosies of the age.

Bernard Lonergan calls religious conversion an "other-worldly falling in love," the giving of the self to the Will of God for the world.[1] It is a love, in other words, that is unlike the love we see around us, grounded on different principles, illumined by a different light, meant to witness to something other than dominance and difference and the idolatry of maleness. Discipleship, that surrender to the presence of Christ in life, presumes from each of us, from the church itself, that

1 Bernard Lonergan, *Method in Theology*, (New York: Herder & Herder, 1972) 237–40.

same kind of reckless, open, receiving, giving love that Jesus brought to the blind on the roads of Galilee, to the body of a dead girl, to the plea of the menstruating woman. Society called the blind sinful, a female child useless, the woman unclean, all of them marginal to the system, condemned to the fringes of life, excluded from the center of the synagogue, barred from the heart of the Temple. But Jesus takes each of them to himself, despite the laws, regardless of the culture, notwithstanding the disapproval of the spiritual notables of the area, and fills them with Himself. To be disciples of Jesus means that we must do the same. There are some things, it seems, that brook no rationalizing for the sake of institutional niceties. Discipleship infers, implies, requires no less than the love of Jesus for everyone, everywhere regardless of who would draw limits around the love of God.

Discipleship and faith are of a piece. To say that we believe God loves the poor, judges in their behalf, wills their deliverance, but do nothing ourselves to free the poor, to hear their pleas, to lift their burdens, to act in their behalf, is an empty faith indeed. To say that God is love and not ourselves love as God loves may well be church but it is not Christianity. To say that all persons are equal in God's sight and ourselves maintain a theology of inequality, a spirituality of domination in the name of God, is to live a lie.

Discipleship recognizes the face of the Divine, the blessing, in what the world calls burden, and unmasks in what we call burdens their blessing to us all. But if discipleship is the following of Jesus, beyond all bounds, at all costs, for the bringing of the Reign of God, for the establishment of right relationships, then to ground a woman's calling to follow Christ in her ability to look like Jesus obstructs the very thing the church is founded to do. It obstructs a woman's ability to follow Christ to the full. And it does so for the sake of religion and in defiance of the gospel itself. How can a church such as this call convincingly to the world in the name of justice to practice a justice it does not practice itself? How is it that the church can call institutions to deal with women as full human beings made in the image of God when their humanity is precisely what the church itself holds against them? In the name of God. It is a philosophical question of immense proportions. It is the question which, like slavery, brings the church to the test.

For the church to be present to the woman's question, to minister to it, to be disciple to it, the church must itself become converted to the issue. In fact, the church must become converted by the issue. Men who do not take the woman's issue seriously may be good priests and kind husbands and caring brothers and faithful friends but they cannot possibly be disciples. They cannot possibly be "other Christs": Not the Christ born of a woman. Not the Christ who commissioned women to preach him. Not the Christ who took faculties from a woman at Cana. Not the Christ who sent women to preach resurrection to apostles who would not believe it. Not the Christ who sent the Holy Spirit on Mary the woman as well as Peter the man.

If this is the Jesus whom we as Christians, as church, are to follow, then the discipleship of the church is now mightily in question.

Indeed, when the poet Basho writes, "I do not seek to follow in the footsteps of those of old. I seek the things they sought," the insight sears the Christian soul as well. Genuine discipleship is not a matter of historical repetition. It depends for its authenticity on our bringing the will of God for humankind to the questions of this age as Jesus did to His. As long as tradition is used to mean following in the footsteps of our past rather than seeking to follow the spirit of Christ, it is unlikely that we will preserve more than the shell of the church. The very consistency of the faith will be shattered beyond repair. Baptism and Eucharist, full for half of us but not for all of us, will be a sham. We will have preserved pieces of the past in the name of discipleship to the detriment of the fullness of Christianity in the present.

Humanity across differences has become the thread that binds the world together in a global age. What was once a hierarchy of humankind is coming to be seen for what it is: the oppression of humankind. The colonization of women is as unacceptable now as the colonial oppression of Africa, the Crusades against Turks, the enslavement of blacks and the decimation of Indians in the name of God. The humanization of the human race is upon us. The only question for the church is whether the humanization of the human race will lead as well to the Christianization of the Christian church. Otherwise, discipleship will die and the integrity of the church with it.

We must take discipleship seriously or we shall leave the church of the future with functionaries, perhaps, but certainly without disciples. The fact is that Christianity lives in Christians, not in books, not in documents, not in platitudes. Discipleship to women and the discipleship of women is key to the discipleship of the church. The questions are clear. The answer may, at this point, seem obscure and uncertain but the conclusion is not merely academic: it is crucial to the future of the church, it is the essence of discipleship, and it is the ultimate measure of Christianity.

Feminist Biblical Interpretation *and* The Renewal *of the* Church

Sandra M. Schneiders, IHM

Introduction

I first met Margaret Brennan, IHM, the woman this volume honors, when I was a nineteen year old novice and she the young assistant novice mistress in the Congregation to which we both belong. The foresight of our Congregation's leaders had allowed Margaret Brennan to earn a Ph.D. in theology in the very first doctoral level theology program ever opened to women, established at St. Mary's College, Notre Dame, Indiana, in 1943. As I sat in her passionately taught courses in theology, and especially in Scripture, I imbibed not only the subject matter and a fascination for Scripture that would inform my own professional choices but, subliminally, a deep conviction that Scripture study was finalized by spirituality and a clear image of scholarship in the person of a woman. It seems appropriate, therefore, to dedicate an essay on the contribution of feminist biblical scholarship to renewal in the Church to one who pioneered as a woman in the biblical field, once the monopoly of men, and who has contributed in a truly remarkable way to the renewal of the Church through her teaching and writing and her deep influence on religious life, especially in North America.

What is Feminist Biblical Interpretation?

Scholars generally acknowledge that the term "feminist biblical interpretation" actually cannot be used in the singular because there are a number of different hermeneutical agendas and procedures that have been developed out of the life experience of different communities of women. Just as there is no such thing as "generic woman" but only individual women in a variety of socio-cultural communities who have something in common as women but who have very different historical experiences of oppression and liberation, so there is no "generic feminist interpretation,"[1] but only Black, Hispanic, Latin American, Native

1 See Elizabeth Schüssler Fiorenza, "Transforming the Legacy of *The Woman's Bible*" in *Searching the Scriptures: A Feminist Introduction*, vol. 1, ed.

American, European, white American, African, and Asian feminist biblical hermeneutics. The different hermeneutical agendas and practices in these communities are responses to diverse experiences of the presence and role of the Bible in diverse believing communities.[2] I write from a particular social location, that of a white, American, Catholic feminist interpreter who makes no claim to universality. However, I trust that what I have to say would be acknowledged and affirmed not only by white feminist interpreters but also by womanist, mujerista, Asian, and African interpreters even if their own perspectives and agendas would be different. Feminist biblical interpreters have much in common without being uniform or interchangeable and we are increasingly aware that no one perspective has priority in defining the agenda.

Although feminist biblical interpretation began over a century ago with the work of women like Elizabeth Cady Stanton, whose book, *The Woman's Bible*, was published in 1894, the history that is of particular interest in this essay, namely, the renewal of the Church, began with Vatican II. For Catholics one of the most invigorating results of the Council was the rediscovery of the Bible as a personal and communal resource for faith after a nearly 400 year scriptural slumber which resulted from the Council of Trent's effort to "protect" Catholics from Protestant notions of private interpretation. Unfortunately, the price of this dubiously valuable protection from misinterpretation was effective denial of deep, personal contact with and engagement of the biblical texts. Most Catholics were abysmally ignorant of Scripture when Vatican II, in "Dei Verbum, the Dogmatic Constitution on Divine Revelation," declared it the "pure and perennial source of the spiritual life."[3] "Sacrosanctum Concilium, the Constitution on the Sacred Liturgy" ordained that Scripture was to be made abundantly available

---

Elisabeth Schüssler Fiorenza with the assistance of Shelly Matthews (New York: Crossroad, 1993): 1–24, esp. pp. 16–21.

2 Part I of *Searching the Scriptures*, vol. 1, contains six essays by women from six different communities of interpretation: E. Gössman on European, K. Baker-Fletcher on Black American, C. De Swarte Gifford on early American, R. Nakashima Brock on Asian American, T. Okure on African, and A. M. Isasi-Diaz on Hispanic interpretation.

3 "Dei Verbum" VI:21. This document is available in English as "Dogmatic Constitution on Divine Revelation," in *Documents of Vatican II*, vol. 1, ed. Austin P. Flannery (Grand Rapids, MI: Eerdmans, 1975): 750–765.

to the People of God, especially in the context of the celebration of the Sacraments. It was to be presented in the vernacular with suitable homiletic elaboration so that all would be nourished at the table of the Word as they were at the table of the Eucharist.[4]

Catholics who took up the Bible in the post-Conciliar period were generally not burdened with the conflicts around biblical authority and especially inerrancy which so exercised Protestants in the wake of the 19th century developments in "higher criticism." Catholics had been educated to regard the teaching office of the Church, i.e., the *Magisterium* (which was functionally equivalent in the Catholic imagination to the Pope or "Rome,") as the ultimate if not the only authority in religious matters. Consequently, what Catholics found in Scripture was not an inarguable authority which had, somehow, to be reconciled with competing scientific data, but a beautiful and challenging story of God's creation of, calling of, journeying with, and finally becoming part of humanity. The enthusiasm of Catholics for this rediscovered spiritual treasure expressed itself in eager participation in Bible study and prayer groups, the choice of biblically based retreats, private biblical prayer and meditation, attendance at Scripture workshops and summer schools, and a revising of preaching and teaching to place salvation history as it is recounted in Scripture at the center of the endeavor to assimilate and hand on the faith. For women whose feminist consciousness was beginning to be raised Scripture presented a particularly consoling possibility, a medium of encounter with God which was not controlled by and doled out to them by clerics. The Bible, unlike the other sacraments, women could "take in their own hands" and relish from within their own feminine experience without permission from or submission to men.

However, this biblical honeymoon was very short-lived! As Catholics began to read the Bible in their own language serious problems presented themselves. What was one to do with the anti-Judaism in the Gospels? with the divinely sanctioned (even commanded) warmongering, land-grabbing, and general violence toward opponents in the Hebrew Scriptures? with the legitimation of the institution and practice of slavery in the Pauline corpus? with the pervasive denigration

4 See "Sacrosanctum Concilium," esp. I, 35; II, 51–52. The latter is available in English as "The Constitution on the Sacred Liturgy," in *Documents of Vatican II*, vol. 1, pp. 1–40.

and marginalization of women? Feminist biblical scholarship which attempted to address some of these problems, especially those which concerned women, not only did not dissipate the problem but deepened it. As scholars approached the problematic texts it quickly became clear that the problems were neither few in number nor superficial in character. They could not be explained away as faulty translations or misunderstandings by commentators. The problems were in the text and they were real.

Even more devastating than the realization that there were problematic texts in the Bible was the growing awareness, finally irrefutably crystallized in Elizabeth Schüssler Fiorenza's ground-breaking book, *In Memory of Her*,[5] that the Bible as a whole is androcentric, i.e., male-centered, and patriarchal, i.e., male-dominative, from end to end. In other words, we realized that the Bible was a book by men, for men, about men, and that "men" is not a generic term. Women generally appear in the Bible, as they have in society for most of history, by male permission, defined by men, for male purposes. Furthermore, the social ordering presumed throughout the Bible is one in which the pattern of domination/subordination, i.e., patriarchal relationship, is ubiquitous and unquestioned. Indeed, this pattern is usually presented as divinely ordained and sanctioned. Thus slavery, destruction of enemies, domination of women and children, and, in the New Testament, anti-Judaism are not chance aberrations but expressions of the ideology which saturates the biblical world. Of course, this is not all that is in the Bible. But the extent of the problem was so overwhelming that many women and men committed to agendas of liberation and justice were simply alienated from the biblical text. Its pervasively non-inclusive language, obsessive androcentrism, oppressive patriarchy, nearly universal male God-imaging, and disturbingly violent sexism made the biblical text no longer part of the solution for many people but part of the problem which was looming large on the Church's agenda within a couple decades of the ending of the Council.

For many women biblical scholars who were as devastated by these unavoidable discoveries as were their fellow believers in the pews, a daunting agenda appeared on the academic horizon. At first we asked the question, "Could the biblical text be saved?" But we quickly realized

[5] Elisabeth Schüssler Fiorenza, *In Memory of Her: A Feminist Theological Reconstruction of Christian Origins* (New York: Crossroad, 1983).

that that was the wrong question. The much more important question is "Can women be saved?" Can they be saved from the oppressing fallout, the subordinationist conclusions, the marginalizing applications to their lives in Church and society of what was in these texts? Could faith survive with such a text as one of its foundations?

Although I cannot trace in any complete or adequate way the history of responses by feminist biblical scholars to this challenging situation,[6] I will mention some typical approaches that have been sometimes problematic but more frequently fruitful, i.e., some contributions of feminist interpretation to the renewal agenda. One agenda item which has been on the table since the beginning is that of translation, specifically the elimination of sexist, or more generally non-inclusive, language so that all members of the People of God can hear and read the biblical text as equals in God's family. But the more we work with this problem the thornier it becomes. The fundamental dilemma which we face is that if we *do not* correct the sexist language in the Bible it can be so offensive to justice-minded people that they refuse to engage the text at all and that if we *do* correct it we risk covering up precisely the problem we need to face and to solve.[7] The Inclusive New Testament,[8] recently published by Priests for Equality, which has literally given the New Testament back to many women, richly illustrates this dilemma. For example, it rewrites Paul's admonition that slaves submit to their masters as to Christ (Eph. 6:5–8) as an encouragement toward coopera-

6 A number of accessible "mappings" of feminist biblical criticism have appeared. See, e.g., Mary Ann Tolbert, "Defining the Problem: The Bible and Feminist Hermeneutics," in *The Bible and Feminist Hermeneutics*, ed. Mary Ann Tolbert [Semeia 28] (Chico, CA: Scholars Press, 1983): 113–126; Carolyn Osiek, "The Feminist and the Bible: Hermeneutical Alternatives," in *Feminist Perspectives on Biblical Scholarship*, ed. Adela Yarbro Collins (Chico, CA: Scholars Press, 1985): 93–105; Janice Capel Anderson, "Mapping Feminist Biblical Criticism: The American Scene, 1983–1990," in *Critical Review of Books in Religion* (Atlanta, GA: Scholars Press, 1991): 21–44, which contains a very extensive bibliography.

7 This problem is very well described by Elizabeth A. Castelli, "Les Belles Infidèles/Fidelity or Feminism? The Meanings of Feminist Biblical Translation," in Schüssler Fiorenza, ed. *Searching the Scriptures*, vol. 1, pp. 189–204.

8 *The Inclusive New Testament* (Brentwood, MD: Priests for Equality, 1994).

tion between employers and employees. The rewriting is certainly more edifying, in the literal sense of building up our faith, for the contemporary Christian, but is it what Scripture says? And do we adequately handle the mistakes in our religious past by writing them out of our records? What we do not remember we are in danger of repeating.

While the problem of language is by no means solved, struggling with it has made some very important contributions to renewal in the Church community. For one thing, it has convinced everyone, oppressors as well as oppressed, that language matters. Indeed, so clear has this become that those determined to maintain the oppression of women in the Church have been forced to draw their "line in the sand" precisely over the language issue. The U.S. Bishops, e.g., insisted on translating even a contemporary document that makes no claim to divine inspiration, namely, the (so-called) Universal Catechism,[9] in the most offensively sexist language possible. This gratuitous linguistic violence testifies to the fact that the oppressors *do* know that this is now an issue. Perhaps the first act in the revolution is to get the attention of the nobility! Or, less facetiously, the first order of business in renewal is to get some control of the agenda. The increasing virulence of the official response to the language issue on the part of those who are determined to maintain the subordinate status of women in the Church is a clear signal that trivializing the language issue is no longer an option. Increasing repression often signals the imminent demise of a regime.

Secondly, the problem of inclusive language is making all of us pay very close attention to what texts do and do not mean and not merely to what words are used. We no longer assume that an address which begins with "Brothers" was issued to an all-male audience much less that contemporary women will hear such an address as including them.

Thirdly, we have been led to consider in depth the diverse uses of Scripture in various settings such as liturgy, personal meditation, and historical critical classroom study. Scripture used in the liturgy is addressed to the whole People of God, not as historical information but as present challenge to transformation. It therefore demands inclusivity in a way that the same texts being studied in a theology classroom might not.

9 *Catechism of the Catholic Church* (Collegeville, MN: Liturgical Press, 1994).

A second approach by feminist scholars to the problematic of the Bible has been serious work on specific texts. Some scholars have devoted their attention to texts which have been used to women's detriment. For example, the creation story in Genesis 2–3 which seems to suggest that woman is derivative from and inferior to man and ultimately responsible for sin has been intensively studied.[10] The subordinationist texts in the Pauline corpus such as 1 Cor. 11:1–16 and 14:34–40 and Eph. 5:22–24 which declare that women are to be subject to men and silent in the Christian assembly have been the object of numerous studies. The object of this work is to undermine the oppressive potential of these texts for contemporary believers. Other scholars have worked on texts which seem to present women in a positive light, such as the stories of the fidelity of women disciples in the Passion and Resurrection accounts,[11] or of the apostolic role of the Samaritan Woman in Jn. 4. Such texts, often misinterpreted or trivialized, suggest that women were much more central to the life of the early Christian community than has generally been admitted. Still other scholars have concentrated on "texts of terror,"[12] horrendous narratives of the violence done to women by men such as the story of the sacrifice of Jephthah's daughter in Jg. 11 or the story of the gang rape, murder, and dismemberment of the concubine in Jg. 19. These texts of terror from our past must become perpetual deterrents to such acts of terror in the present and the future.

Again, there is ambiguity in this work. Concentrating on stories about women can appear to be an acceptance of the wrong presuppositions. For example, it can suggest that only passages in which women appear or are specifically dealt with pertain to women whereas the whole of Scripture, by common consent, pertains to men. Or it can suggest that women in Scripture are models only for women whereas men are models for everyone. Nevertheless, these endeavors have acquainted the People of God with women figures in Scripture who have been largely ignored throughout the liturgical and theological history of the Church, for example, the courageous and intelligent midwives, Shiphra

10 Gen. 2:21–23 suggests to some the derivative nature of woman in relation to man and Gen. 3:1–6 seems to make woman responsible for the Fall.

11 E.g., Mt. 28:1–11; Mk. 16:9–11; Jn. 20:11–18.

12 This expression was coined by Phyllis Trible in her book, *Texts of Terror: Literary-Feminist Reading of Biblical Narratives* (Philadelphia: Fortress, 1984).

and Puah, who saved the Hebrew race from the genocidal plot of the king of Egypt (Ex. 1:15–22),[13] Hulda the prophet of the Josian reform (2 Kgs. 22:14–20), Phoebe, the deacon of the early Christian Church in Cenchrae (Rom. 16:1–2), Prisca the evangelist (Acts 18:26), and Junia the apostle (Rom. 16:7). These studies have undercut the seemingly self-evident proposition that women were of minor significance in the Old Testament and early Christianity, and they have called into question the tacit acceptance or whitewashing of biblical violence against women. Studies of particular texts have also opened up the possibility that, looked at through feminist eyes, such texts do not necessarily support the patriarchal agenda. Perhaps, e.g., the creation of Eve from the comatose Adam does not support the hypothesis that woman is inferior to man any more than the creation of Adam from the inert dust suggests such a relationship of Adam to the mud! Perhaps the tacit assumption that the companion of Clopas on the road to Emmaus was another male rather than (as would seem more natural) Clopas' wife says more about the sexual imagination of celibate male interpreters than about discipleship in the early Church.

A third type of feminist biblical interpretation is specifically theological in nature and consists in re-investigating the proposed biblical bases of Church positions that are oppressive of women. Is it the case, for example, that the refusal to ordain women rests on inarguable biblical data?[14] The Pontifical Biblical Commission which was assigned by the Vatican to study this question (and, no doubt, expected to conclude in favor of the exclusion of women) came out with the "wrong" conclusion in its 1976 "report" which was leaked to the press despite Vatican efforts to keep it secret. The Commission voted that the Bible does not supply the grounds for barring women from ordained ministry, that it does not settle the question of ordination one way or the other, and that

13 My colleague John Endres, in an unpublished paper, did a wonderful interpretation of this passage which called it to my attention.

14 This is the claim made in the "Responsum" of the Congregation for the Doctrine of the Faith issued on October 28, 1995 and made public on November 28, 1995: "This teaching [that the Church has no authority whatsoever to confer priestly ordination on women] requires definitive assent, since, founded on the written Word of God...." The responsum was a clarification of the teaching of Pope John Paul II in "Ordinatio Sacerdotalis," issued in May 1994.

ordaining women would not be contrary to the will of Christ as it can be discerned from the New Testament data.[15] Meanwhile, a task force of women and men feminist scholars in the Catholic Biblical Association was working on the same question and concluded in its 1979 report that "the NT evidence, while not decisive by itself, points toward the admission of women to priestly ministry."[16]

While this type of work may seem futile in that the conclusions by teams of reliable biblical scholars have had no effect on the minds of those who have the power to make institutional changes in the Church, it is actually very important. Such studies have not only abolished the theological persuasiveness of the arguments against the ordination of women but also undermined the capacity of ecclesiastical officials to intimidate the people in the pews into accepting as literally "Gospel truth" anything in an official document which has a biblical citation in parentheses after it. Catholics have learned to check the references and to question the interpretation of biblical data, no matter who is doing the interpreting, especially when the interpretations favor oppression. This healthy suspicion regarding the imputed biblical basis of dogmatic positions has been applied also to such issues as mandatory clerical celibacy, the prohibition of contraception, condemnations of homosexuality, and so on. It is, in other words, dangerous to teach the oppressed to read, especially if they start reading the fine print!

A qualitative leap forward in feminist biblical interpretation occurred with the publication in 1983 of Elizabeth Schüssler Fiorenza's book, *In Memory of Her*. This volume did not work on words, individual texts, or specific problems. Fiorenza undertook a full scale feminist critical commentary on the entire New Testament. Beginning with a global hermeneutical suspicion of the pervasive androcentric and patriarchal character of the text as a whole, Fiorenza worked her way through every book of the New Testament with the avowed objective of restoring women to the center of Christian history and their

15 For a full treatment of this document and its significance, see John R. Donahue, "A Tale of Two Documents," in *Women Priests: A Catholic Commentary on the Vatican Declaration*, ed. Leonard Swidler and Arlene Swidler (New York: Paulist, 1977): 25–34.

16 "Women and Priestly Ministry: The New Testament Evidence," a report by the Task Force on the Role of Women in Early Christianity, *The Catholic Biblical Quarterly* 41 (October 1979): 608–613.

Christian history to the center of women's self-understanding.[17] Her work rocked the biblical *status quo* and established the absolute necessity of taking the feminist agenda seriously within the theological and biblical academy. Whether or not one agreed with every strategic move or interpretive conclusion of Fiorenza's work there was no avoiding its global implications. And in the process of carrying out the project Fiorenza developed methods of dealing with the biblical text which other feminist scholars have continued to utilize and refine.[18]

A number of other feminist scholars have developed interpretive approaches which, like that of Fiorenza, attempt to deal consistently with the whole of the biblical text from a liberating hermeneutical perspective. Phyllis Trible in her wonderful book, *God and the Rhetoric of Sexuality*,[19] traces the theme of relationship between man and woman as image of God throughout the whole of the Hebrew Scriptures from the Creation account in Genesis to the apotheosis of mutual love in the Song of Songs, in order to unearth a potentially subversive egalitarian mutuality underlying the more obvious patriarchalism on the surface of the text. Letty Russell has used, as a global hermeneutical principle, God's original plan for humanity as it is voiced not in what already is but in the groaning of the Spirit of God in our spirit toward a New Creation.[20] The Bible, she has maintained, is not closed, but open-ended. It leads us to hope for, and therefore to move toward, that which it does not and cannot fully and explicitly describe but toward which it points, namely, the fulfillment of God's promise of liberation, justice, mutuality, universal *shalom*. And Rosemary Radford Ruether has used the Old Testament prophetic protest against all injustice, all dehumanizing of persons by the established powers, as a principle of discernment of what in the biblical text is life-giving and what is not, i.e. what is true to the biblical message and what is a falling away, within the text, from the Bible's own message. The Bible, in her work, becomes its own critic, dis-

17 Schüssler Fiorenza, *In Memory of Her*, xx.

18 See Schüssler Fiorenza, *In Memory of Her*, ch. 2.

19 Phyllis Trible, *God and the Rhetoric of Sexuality* (Philadelphia: Fortress, 1978).

20 See, e.g., Letty M. Russell, "Authority and the Challenge of Feminist Interpretation," in *Feminist Interpretation of the Bible*, ed. Letty M. Russell (Philadelphia: Westminster, 1985): 137–146.

allowing that within itself which is contrary to the full humanity of all God's people.[21]

In my own work I have attempted to deal with still a different aspect of the problematic, namely, how we can understand the theological, literary, and historical nature of the biblical text as text in such a way that we can engage it dialogically. Is it possible, on the one hand, to refuse to submit totally and blindly to the content of oppressive texts and, on the other hand, to refuse to dominate the text by denying it any authority which we do not confer upon it, i.e. by selecting the texts to which we will grant a voice and declaring others non-revelatory? In other words, I have been searching for an interactive or dialogical hermeneutical model which can allow both reader and text to change and grow through the interaction, the encounter, that we call reading.[22]

One implication of such a theory is that we must take with utmost seriousness the *humanity* of the text as well as its divinity just as we must take utterly seriously the humanity of Jesus. As human witness to divine self-revelation the biblical text is limited, fallible, and sometimes simply wrong. But just as we do not reject totally people who are not perfect, but engage them in mutually clarifying dialogue which ideally changes both of us, so we need not reject the Bible as human text because it is not, in every respect, the expression of all that humanity has learned over the past 2000 years. This text, limited as it is, is human witness to *divine* self-revelation and as such it is a potentially precious mediation of encounter with God. The task is to discover a way to interpret the text which will allow us to refuse its errors and evils without destroying its capacity to mediate God in our lives.

21 See, e.g., Rosemary Radford Ruether, "Feminist Interpretation: A Method of Correlation," in *Feminist Interpretation of the Bible*, 111–124. A remarkable confirmation of this approach, from a completely unrelated source, is supplied by Gil Baillie in *Violence Unveiled: Humanity at the Crossroads* (New York: Crossroad, 1995), 199, where he states: "The God depicted in the Bible is not always synonymous with 'the biblical God,' and sometimes the two are profoundly incompatible. When the former is vengeful and violent, the latter begins undermining the myths of sacred violence that endow such violence with religious legitimacy. The struggle between these two religious realities is what the Bible is all about...."

22 I have developed this theory in *The Revelatory Text: Interpreting the New Testament as Sacred Scripture, 2nd ed.* (Collegeville, MN: Liturgical Press, 1999).

A second implication, therefore, is that we cannot escape the work of interpretation and that entails wrestling with the text, as Jacob wrestled with the angel, until we obtain a blessing. Elsewhere I have written about some techniques, some "holds" as it were, in this wrestling match. For example, rather then asking simply what a text says and then either submitting to it or dismissing it as non-revelatory, we need to ask what kind of trajectory the text opens up, what newness it introduces? When Paul commands, "Wives be subject to your husbands, and husbands love your wives" the newness is not in the repetition of the societal injunction to wives. That belongs to the stock in trade of patriarchal cultures. What is new is the injunction to husbands to *love* their wives. The text remains patriarchal and problematic but it also opens up a path, a trajectory—namely, love—which, if followed out, must sooner or later call into question unilateral subordination. Love and domination are finally incompatible. Paul is not only underwriting his society's patriarchal flaws (which is lamentable) but setting Christians on a journey which must someday subvert that very patriarchy.

We must in my view develop and use a variety of interpretive techniques such as this one, which will allow us to use the whole of Scripture for our personal and communal spirituality. Sometimes, as we wrestle with the text, it will mirror to us the darkness in our own hearts, in our Church, and in our society, and challenge us to repudiate that darkness not only in the text but in and among ourselves. At other times it will illuminate us with truth and energize us with love. Just as in our own personal histories we grow both from our mistakes and our successes, just as our sins can be as revelatory as our virtues and sometimes more so, so the scriptural text, if wrestled with creatively and imaginatively, can in its wholeness be a blessing. We will not escape this struggle without scars but when we emerge from it we will know that we have been, even in the dark, in the mysterious and life-giving presence of God. By the same token, the text will also be different, no longer secure in unquestioned and oppressive domination in the name of God, but made supple and open to new meaning by its interaction with the experience of the readers who do not share or accept the oppressive ideology which presided over the text's development. The materiality of the words in the text does not change in the process of interpretation, but the relation which we call meaning does.[23]

23 I have discussed in some detail some of the methods of struggling fruitful-

All of this feminist biblical work, and much that I cannot mention here, has led to some radical questioning of assumptions about the very nature of the Bible that were, until a few decades ago, simply taken for granted. Feminist scholars have been especially exercised by the question of biblical authority. What kind of authority does the Bible have in the faith and life of the believer? Some feminist scholars have tended to deny authority to the text itself and to locate authority in the community of women and men struggling for liberation whose task, they say, is to discern what in the biblical text is genuinely liberating and life-giving and to confer authority upon these and only these texts.[24] I do not think, for reasons just articulated, that this is the most helpful way to formulate or to deal with the question of biblical authority. First, it suggests that authority is a kind of objective quality inhering in either texts or people rather than a characteristic of authentic relationships. Second, it suggests that authority equals dominative power, i.e., that texts with authority must be submitted to and that texts to which we cannot submit must be denied authority. Nevertheless, even if not fully adequate, this way of formulating a response to questions of biblical authority has the advantage of helping to liberate us from an idolatry of the biblical text considered as an objective content which finally confronts the Christian with the choice of either submitting to what is dehumanizing or leaving the Christian community to those who are willing to submit. It offers us the challenge to find a way of understanding the role of the Bible in Christian faith that is analogous to the challenge of finding a way to deal with the authority of officials in the Church and with the authority of documents which testify to our theological tradition. Neither radical rejection nor slavish submission is a freely human and Christian response. Blind submission on the one hand or denial of authority on the other cannot be the only possibilities. What we need is a reformulation of our understanding of the nature and function of authority.

---

ly to interpret problematic texts in a pamphlet entitled, *How to Read the Bible Prayerfully*, reprinted by Liturgical Press from "God's Word for God's People," *The Bible Today* 22 (March, 1984): 100–106.

24 See, e.g., Elisabeth Schüssler Fiorenza, *Bread Not Stone: The Challenge of Feminist Biblical Interpretation* (Boston: Beacon, 1984), 145, among other places.

The work of feminist scholars has also problematized the canon of Scripture, that is the official list of biblical books. Scholars have brought out into the light many texts, composed during the same period in which the New Testament materials were written, which did not make it into the New Testament canon. They have asked why these texts, many of which tell the stories of women leaders in the early Church, were rejected and marginalized.[25] The traditional explanation for the non-canonical or apocryphal status of these works is that they did not pass the test of orthodoxy or true faith in the early Christian community. But we have learned to be much more critical of such simplistic explanations. "Heresy" is the description of the historical losers by the historical winners and it may or may not be strictly equitable with false teaching. While it may not be likely that the New Testament canon will be expanded to include more than the twenty-seven books it now includes we may very well learn to read these twenty-seven in the context of the whole early Christian corpus the way we read the Constitution in the light of The Federalist Papers or, even more instructively, the way Jews read the Torah in the context of Mishnah and Talmud.[26] In other words, canonical does not necessarily denote that which is orthodox and true and non-canonical that which is heretical and false. And many works written after the time of biblical composition may be more important for our faith than some materials in the Bible. If one were to compare the Epistle to Titus and Teresa of Avila's Interior Castle in terms of their spiritual wealth and transformative potential the latter should probably be chosen over the former. In other words, canonical imprisonment is no more objectively valid than textual idolatry. The relation of the believer to Scripture, especially to the New Testament, must have the subtlety, flexibility, and nuance of an adult relationship rather than the rigidity of childish literalism.

[25] For these texts, see *New Testament Apocrypha*, ed. Wilhelm Schneemelcher, Eng. tr. by A. J. B. Higgins et al., ed. by R. McL. Wilson (Philadelphia: Westminster, 1963–1966), 2. vols.

[26] For a clear presentation of the much more inclusive Jewish approach to text and "Scripture," see Wilfred Cantwell Smith, *What Is Scripture? A Comparative Approach*.(Minneapolis: Fortress, 1993), ch. 5.

What has Feminist Biblical Hermeneutics Achieved?

After this brief and selective sampling of the work of feminist biblical scholarship I would like to suggest some global contributions it has made to the overall renewal agenda in the Church. I think these contributions can be divided into two general categories: theoretical and programmatic contributions on the one hand, and substantive or content contributions on the other.

THEORETICAL AND PROGRAMMATIC CONTRIBUTIONS

A first programmatic contribution of the intensive scrutiny of the biblical text by feminist scholars is that it has problematized, that is, put in question, certain previously taken for granted misunderstandings about the nature and properties of Sacred Scripture. As already suggested, it has called into question the meaning of canon (understood as designating these and only these books as valid witnesses to early Christian faith and life), biblical revelation (understood as meaning divine dictation), biblical inspiration (implying absolute infallibility of texts), and biblical authority (meaning textual domination of faith). The meaning of all of these properties or qualifications of Scripture have to be nuanced and even reformulated. Questioning the absolutist conception of "God's will" implied in such simplistic understandings has helped believers "take back the text" as mediation of the encounter with God. It is increasingly difficult for people in authority, whether ecclesiastical officials or political right wingers, to use the biblical text as a kind of bludgeon to coerce the submission of believers.

Second, feminist and other liberationist biblical interpreters have been teaching people, especially oppressed people, to read the Bible from the standpoint of their own experience and to trust their reading as genuinely authoritative. Among first world women the publication of "Inter Insigniores," the document denying ordination to women, was a watershed event in this domain. The document tried to tell women that the Bible was the ultimate basis of their second class citizenship in the Church.[27] Refusing this reading was a significant step in women's "tak-

[27] "Inter Insigniores," 2:10–13. An English translation of the document is available as "Declaration on the Question of the Admission of Women to the Ministerial Priesthood," in *Women Priests*, 37–49. This position has been

ing back their lives" in relation to the authority of both Scripture and Church officials. In learning to unmask oppressive material in Scripture and distinguish patriarchal and sexist projections in the text from "God's will" many are learning to do the same thing with magisterial pronouncements and Church documents. Difficult as it is, against the background of pre-conciliar Catholic formation to blind obedience, Catholics are coming to the realization that false things do not become true because they are in the Bible or because they are said by someone in office. Authority is a quality of truth; what is untrue is not authoritative no matter who says it.

Substantive or Content Contributions

Other contributions of feminist interpretation to the renewal of the Church consist in substantive theological positions which such interpretation has enabled and underwritten. First, a number of realizations about women in themselves and in their relationship with God have emerged from this work. By raising to visibility numerous positive, and often complex, female figures whose presence in the Bible has been ignored or repressed for centuries, feminist scholarship has been disputing the stereotyping of women in terms of their sexuality as either "good mothers" or whores. It has challenged the tendency to see women as valuable only because of their relationship to men as wives, daughters, mothers, or sisters. It has named and rejected the violence against women often depicted as normal or legitimate in the biblical texts and laid claim to biblical denunciations of male abuse which have often gone "unnoticed." Gradually, feminist scholarship is undermining the biblical legitimization of women's inferiority, subordination, and victimization.

Second, as already mentioned, feminist biblical scholarship has undermined the biblical arguments against the ordination of women and therefore for their relegation to second-class status in the Church, and provided good biblical warrant for women's ordination. It has thus exposed the refusal to ordain women for what it really is. It is not an unfortunately painful but necessary execution of the mysterious divine

---

reiterated in "Ordinatio Sacerdotalis" in 1994 and in the clarifying "Responsum" in 1995.

will but the completely unmysterious power agenda of some churchmen.

Third, especially in the last decade, an enormous amount of scholarly effort has been poured into the exploration of feminine God-images in Scripture and especially the powerful gestalt of Woman Wisdom in the Old Testament, the most fully developed personification of God in the whole Bible[28] whom we have begun to see as the particular divine identity of Jesus, the Wisdom of God made flesh.

Not only has this feminine God-image enhanced women's sense of being truly made in the image and likeness of God, but it has led to a fourth substantive contribution, namely, a re-examination of the biblical image of Jesus, so long presented as the ideal priest, revelation of God's patriarchal masculinity, and incarnate explanation of women's religious and spiritual inferiority. Jesus, Sophia incarnate, is coming to be seen by many as the de-legitimator of male supremacy, the one who in his very person and relationship to God and to his fellow human beings de-stabilizes patriarchy and initiates a discipleship of equals. The risen Christ of the Gospels and especially of the Pauline corpus is the principle of his body, the Church, which is equally female and male, black and white and brown and yellow and red, poor and rich, gay and straight, sick and well, slave and free, Semite and Gentile, inclusively human rather than exclusively male.[29]

28 Biblical work of scholars like Johanna W. H. van Wijk-Bos, *Reimagining God: The Case for Scriptural Diversity* (Louisville, KY: Westminster, 1995) and Susan Cady, Marian Ronan, and Hal Taussig, *Sophia: The Future of Feminist Spirituality* (San Francisco: Harper and Row, 1986) has been strongly abetted by the work of feminist theologians such as Elizabeth A. Johnson, *She Who Is: The Mystery of God in Feminist Theological Discourse* (New York: Crossroad, 1992) and *Women, Earth, and Creator Spirit* [1993 Madeleva Lecture in Spirituality] (New York/Mahwah, NJ: Paulist, 1993), Denis Edwards, *Jesus the Wisdom of God: An Ecological Theology* [Ecology and Justice Series] (Maryknoll, NY: Orbis, 1995), and Sallie McFague, *Models of God: Theology for an Ecological, Nuclear Age* (Philadelphia: Fortress, 1987).

29 I have elaborated this thesis more fully in *Women and the Word: The Gender of God in the New Testament and the Spirituality of Women* [1986 Madeleva Lecture in Spirituality] (New York/Mahwah, NJ: Paulist, 1986).

The figure of Mary is also emerging from feminist biblical work as the powerful singer of divine liberation of the lowly and the poor in the Magnificat, the thoughtful and free cooperator with God in the work of salvation, the worthy successor of those great women of the Old Testament like Eve, Miriam, Judith, Esther, and Ruth, rather than as the dainty but unattainable projection of male fantasies whose primary function is to convince real women that no matter what they do they will never measure up to the ideal of virginal motherhood and to assure them that the less they say and the more they suffer the holier they are.

Another area of substantive contribution to renewal emerging from feminist biblical scholarship is ever increasing light on the divine plan of salvation. This may be the most fundamental and far-reaching result of the feminist re-shaping of biblical interpretation. Feminist scholarship has irreversibly called into question the claim of patriarchy and ultimately of hierarchy itself to divine legitimation. Not only is it becoming increasingly clear that patriarchy in particular and hierarchy in general is not necessarily the only divinely approved form of social organization in family, Church, and society but also that these essentially dominative power arrangements may be so contrary to God's will for humanity, no matter how ancient and sacralized they may be, that they will have to be finally abandoned.

Feminist scholarship is supplying believers and facing the institutional Church with some radical questions. Did *God* create some people intrinsically superior to others, e.g., men to women, whites to people of color, heterosexuals to homosexuals, or is this the invention of the socially powerful for the domination of the threatening "other"? Did *Jesus* found a power structure or did he call into being a discipleship of equals whose appropriate forms of self-organization must always reflect the mutuality and co-responsibility of a Trinitarian God and never degenerate into power-over arrangements such as we find, Jesus says, among the pagans (cf. Mk. 10:42–45)? What are the implications of the answers we give to such questions for maintaining that the *Church* is "hierarchical by divine institution"? Does this oft-repeated maxim perhaps belong to the same category as "slaves be subject to your masters" (cf. Eph. 6:5) or "women must keep silence in the churches" (cf. 1 Cor. 14:34)?

Conclusion

It is probably premature to guess whether, in the last analysis, feminist biblical interpretation will be able to save the Bible for women or to save women from the Bible. The problems with the text are enormous and, for many people, the time is too short. What is being done, however valuable, is for them too little, too late. Nevertheless, for many justice minded people, women and men, the Bible has played too important a role in their spirituality, has offered too much encouragement in their struggles, has mediated Jesus and his mysterious Divine Origin too deeply and tenderly, has marked too indelibly their sense of religious identity and their ministerial commitment for them to give up on this "perennial source of the spiritual life." For better or for worse they are people of the Word and that Word is living and active (cf. Heb. 4:12) and cannot finally be bound (cf. 2 Tim. 2:9) even by the ancient chains of patriarchy. For these people the promise of Jesus continues to sustain their hope: "If you remain in my word you will become my disciples and you will know the truth and the truth will set you free" (Jn. 8:31).

Women *and* Ministry *in the* Roman Catholic Church

Mary Ellen Sheehan, IHM

The situation of women and ministry in the Roman Catholic church today is marked by change that has accrued over the last thirty years and by challenge that promises to continue well into the future. In the period before Vatican II, the New Testament term *ministry* was not used for the actions of the church carried out by reason of its nature and mission. What prevailed instead were services rendered through a well established pattern that distinguished classes or states in life, arranged hierarchically, each of which was entrusted with carrying out aspects of the church's life and ministry. Clergy, official public ministers by definition, were empowered to exercise all forms of service, with exclusive authority for preaching, sacraments, administration, canon law, and theology. Members of active religious congregations, most of whom were women, provided in a quasi-official way a variety of services, such as catechesis and religious education in missionary, school, college, and university institutions; service to the sick and dying in clinics and hospitals; and care of the poor and needy in various forms of social work. The laity, although called to share in the apostolate of the hierarchy through Catholic Action, the Sodality, and other such forms of action-oriented faith, were largely more the recipients of ecclesiastical service than its agents.

Today, while the hierarchical structure remains, the situation has changed dramatically. The biblical word *ministry* is used widely to describe the church's diverse services and all the members of Christ's faithful are called to engage actively in ministry, according to their baptismal call and Spirit-given gifts. Increasingly, missionary proclamation, catechetical formation, liturgical and sacramental ministry, pastoral care, parish and diocesan administration, and many forms of societal ministry are exercised by the faithful, among whom the largest number are women. As well, the academic ministries of theology and canon law are carried out by both lay and clerical members of the faithful, among whom many outstanding teachers and scholars are women.[1]

1 For an excellent review of scholarship produced over the last thirty years by

In short, the traditional pattern of church ministry is shifting. New forms of ministry have emerged as members of Christ's faithful respond to an awakened awareness of their call to serve the church and society.[2]

While not intending to be exclusive of lay men, this essay is focused on women who are responding to their responsibilities as agents in the church's mission.[3] First, I will identify some of the theological and social foundations for this change. Second, I will probe the relatively new pastoral practice of women as pastoral administrators of parishes for its ecclesiological significance. Third, I will identify some theological issues arising from the current situation of the ministry of women in the Roman Catholic church. With respect to the specific perspective of this essay, I will accent the North American situation since that is the context with which I am most familiar.

---

women who are theologians, see Elisabeth A. Johnson, CSJ, Susan A. Ross, and Mary Catherine Hilkert, OP, "Feminist Theology: A Review of Literature," *Theological Studies* 56 (1995) 327–352. For examples of work produced by women who are canon lawyers, see Sharon Euart, RSM, "Women and the 1983 Code of Canon Law," *Origins* 20 (1990–91) 452–456 and other articles by women in recent volumes of canon law journals.

2 Jay P. Dolan, R. Scott Appleby, Patricia Byrne, and Debra Campbell, *Transforming Parish Ministry: The Changing Roles of Catholic Clergy, Laity, and Women Religious* (New York: Crossroad, 1989).

3 The outpouring of literature on this subject in recent years is astonishing. For some examples, see: L. Baroni, Y. Bergeron, P. Daviau, M. Laguë, *Voix de Femmes; Voies de Passage: Pratiques pastorales et enjeux ecclésiaux* (Montréal: Paulines, 1995); Elisabeth Schüssler-Fiorenza, *Discipleship of Equals: A Critical Feminist ekklesia-logy of Liberation* (New York: Crossroad, 1993); *LCWR Ministry Survey* (Silver Spring, MD: Leadership Conference of Women Religious, 1991); Sarah Bélanger, *Les Soutanes Roses: Portrait du personnel pastoral féminin au Québec* (Montréal: Bellarmin, 1988); World Union of Catholic Women's Organizations, "A Report on Women and the Church," *Origins* 14 (1984–1985) 750–756; Elisabeth J. Lacelle, "Le mouvement des femmes dans les Eglises nord-americaines," *Etudes* 363 (1985) 541–554; Francine Cardman, "'The Church Would Look Foolish Without Them': Women and Laity since Vatican II," in Gerald M. Fagin, SJ, ed., *Vatican II: Open Questions and New Horizons* (Wilmington: DE: Glazier, 1984) 105–133; Rosemary Radford Ruether, *Sexism and God-Talk: Toward a Feminist Theology* (Boston: Beacon Press, 1983), esp. chapter 8.

Foundations for Change in Vatican II and the Women's Movement

Vatican II did not identify the ministry of women as theologians, canon lawyers, and pastoral workers in the church as one of its themes for explicit theological reflection. As is well known, women were just barely recognized at all at this historic event.[4] Nevertheless, developments over the last thirty years have shown that the Council has contributed significantly to the emergence of women in church ministry.[5] The Council's focus on the church as mystery, communion, people of God, and as an institution always subject to reform[6] opened up the church for wider and active participation on the part of all its members. In its stirring proclamations concerning the call of everyone to holiness, the Council challenged anew all the members to serve the church and society. Its teachings on seeking justice as constitutive to the Christian life charged its members to undertake new forms of social analysis and action. Living out the saving grace of baptism and confirmation, the Council taught, calls all of us into renewed discipleship, no longer only

4 Mary Luke Tobin, "Women in the Church: Vatican II and After," *Ecumenical Review* 37 (1985) 295–305.

5 In 1967, inspired by the Conciliar vision, Margaret Brennan, as president of the Immaculate Heart of Mary Sisters of Monroe, Michigan, called the first group of twelve women from among the Monroe congregation to study internationally for doctorates in theology and canon law. As well, she enabled the training and placing of women in pastoral catechetics, parish-based ministry, and basic Christian community ministry in the United States, Puerto Rico, and Brazil, all directed toward an appropriation of the Vatican II vision of church. I have personal knowledge of at least five other religious congregations of women who acted similarly at this time. For some further examples of the emergence of women as theologians and pastoral workers, see: Catherine Mowry LaCugna, "Catholic Women as Ministers and Theologians," America 167 (1992) 238–248; Mary Daly, *Outercourse: The Be-dazzling Voyage* (San Francisco: Harper, 1992) 55–80.

6 As mystery, in the words of Paul VI, the church " ... is a reality imbued with the hidden presence of God. It lies, therefore, within the very nature of the church to be always open to new and greater exploration." Address at the opening of the Council's second session (1963), as cited in Walter M. Abbott, SJ, ed., *The Documents of Vatican II* (New York: America Press, 1966) 15.

for personal holiness but also for the building up of the church and society.[7]

Corresponding to this wider understanding of the church and the social dimension of our faith, the council set in motion a renewed understanding of ministry. While maintaining the traditional hierarchical distinctions of the *sacred* realm of the clergy, the *secular* realm of the laity, and the *mixed* realm of members of religious orders and congregations, the Council called for renewal to occur within each of these groups. The clergy were urged to prune themselves of privilege and status and reclaim the service orientation of ordination. The laity were challenged to move away from being passive receivers and to renew the active character of their faith as Christians. Members of religious orders and congregations were counseled to return to the spirit of their founders and orient themselves anew to ministries of justice and reconciliation.

What in fact has resulted in the ensuing years is a certain blurring of the *sacred-secular* distinction. While clergy still maintain exclusively the *sacred* realm of church authority and the administration of the sacraments, many of them also function in the *secular* realm as scholars and activists, pursuing justice and peace in political and economic life. While members of the laity still function mostly in the *secular* realm through discipleship expressed in their personal, family, and work commitments, many of them also function, at both the parish and diocesan levels, in services related to the *sacred* as lectors, preachers, teachers, evangelists, catechists, agents of sacramental formation, pastoral care givers, chaplains, spiritual directors, administrators, canon lawyers, and theologians. While members of religious orders and congregations maintain their quasi-official, *sacred* status in the church, many of them serve in the *secular* realm to care for the poor directly and to labor for social transformation in societal institutions. Thus, the current pattern of ministry, at least in the North American setting, manifests a more

[7] For the Council's teachings on the church, justice, and the vocation of the laity, see: *Dogmatic Constitution on the Church*, *Pastoral Constitution on the Church in the Modern World*, and *Decree on the Apostolate of Lay People* in Austin Flannery, OP, ed., *Vatican Council II: The Conciliar and Post Conciliar Documents* (Northport, NY: Costello Publishing Co., 1987, study ed.). References to Vatican II documents in this essay are taken from this edition and cited by title and section number.

complex and dynamic pattern in which it is estimated that between 70–85% of church ministers are women.[8]

Another factor influencing the growing presence of women in church ministry is the women's movement which, whether intended or not by the shapers of Vatican II, was destined to intersect with ecclesial renewal. The Council proclaimed the church's need to engage with the social realities of our times, one of which was the changing position of women recognized by John XXIII as a "sign of the times."[9] The Council also taught that all forms of discrimination "... whether based on sex, race, color, social conditions, language, or religion, must be curbed and eradicated as incompatible with God's design."[10] As well, it fostered ecumenical dialogue which brought the Roman Catholic church into contact with churches already engaged in overcoming discrimination by extending the practice of ministry in all its forms to women. When blended with the justice oriented agenda of the women's movement, these Vatican II teachings and ecumenical dialogues were destined to rebound into the church itself in its teaching authority and as an institution.

8 As reported in Bélanger, *Les Soutaines Roses*, pp. 77–97, over 70% of paid pastoral workers in Québec are women, about half of whom are members of religious congregations and half of whom are single or married women. As reported by Philip Murnion in a summary published in *Origins* 21 (1991–92) 777–780 of a study commissioned by the U. S. Bishops Pastoral Research and Practice Committee entitled *New Parish Ministers: Laity and Religious on Parish Staffs*, 85% of ministers in about 9,500 U. S. parishes are women. Four out of ten of these women are members of religious congregations.

9 He acknowledged three such signs: workers rights, women's rights, and global political self-determination. About women, he writes: "...the part that women are now playing in political life is everywhere evident. This is a development that is perhaps of swifter growth among Christian nations, but it is also happening extensively, if more slowly, among nations that are heirs to different traditions and imbued with a different culture. Women are gaining an increasing awareness of their natural dignity. Far from being content with a purely passive role or allowing themselves to be regarded as a kind of instrument, they are demanding both in domestic and in public life the rights and duties which belong to them as human persons." See "Pacem in Terris," no. 41 in *The Encyclicals and Other Messages of John XXIII* (Washington: The Pope Speaks Press, 1964).

10 *Pastoral Constitution on the Church in the Modern World*, no. 29.

While continuing to evolve in complex and self-critical ways, the women's movement remains centered on its two-fold goal of critical social analysis and creative reconstruction.[11] It persists in carrying out the analysis of the oppression of women through pervasive sexism in attitudes, behaviors, and structures, both in society and the church. It continues to show the inadequacy of defining women by roles rather than by their own agency as fully human persons. It perseveres in restoring the memory of women to Christian history. It commits to generating new biblical and theological knowledge and transformational social and ecclesial practice. As such, it is sowing seeds, and even bearing some fruit, for ecclesiological reform.

Advances in feminist theory and theology raise important issues for the church regarding its teaching on the inclusion of women in church ministry. One is the just use of church resources for the admission of women to theological and pastoral training for ministry since currently more support is given to candidates for ordained ministry. Another is the need for ecclesiastical structures to serve the discernment of women's gifts for ministry and to guarantee just and equitable employment conditions. A third is the need to transform the subordination pattern of ministry into a collaborative one in which women work as genuine and equal partners with the clergy. These issues revolve around how justice functions in a hierarchical ordering of ministry where authority regarding finances, decision-making, and the limitations placed on the use of women's gifts in the church is still linked juridically with ordination. While recent church documents ring out with declarations regarding the equality of women in the church and society, there is still the concrete task of implementing this proclamation in church structures and practices, especially those pertaining to women in its ministerial life.

The overarching issue regarding women in ministry, of course, is their continued exclusion from ordained ministry, a practice which official church teaching at the present time does not regard as a matter of justice but rather as the intended will of God. This exclusion revolves around a few central traditional teachings. Women cannot be ordained because, if it were God's will, Jesus would have ordained them. Women cannot exercise official public leadership with authority since headship

11 Johnson, Ross, and Hilkert, "Feminist Theology: A Review of Literature."

as a symbolic function rests only with men in continuity with Jesus Christ, the true head of the body of believers. Women cannot celebrate the Eucharist, the sacrament of reconciliation, and the sacrament of the sick in the name of the church since only males, by reason of their sex, can be fitting signs to mediate the presence and action of Christ in sacramental action.

For the most part, proponents of the ordination of women acknowledge that ordination is not a human right but rather a gift from God.[12] What they do point out, however, is that theology and church practice are not static. They grow from interaction with new historical analysis, new knowledge, and new pastoral needs. They see that the critical analysis of sexism and gender-based reductionistic thinking contributes to breaking down previously unchallenged patterns of thinking and acting. They also propose that the rapidly developing teaching on the full and equal humanity of women before God opens new doors and creates new horizons from which to contemplate the mystery of God and its expression in Christ's church as a sacrament, the sign and instrument, as Vatican II says, of communion with God and of unity among human kind.[13] It might even be the case that the church must come to see the ordination of women as *necessary*, by reason of their full dignity in God and their fittingness as signs of Christ's salvific action among us, if the church is to be fully a sacrament of God's presence among us.

The Ministry of Women as Parish Administrators

Among the many forms of church ministry currently exercised by women, it is particularly interesting to study their situation as episco-

12 Few would argue that admission to ordained ministry is a human right, although it must be admitted that much popular contemporary discussion is cast from this perspective. It is hard not to see, however, that there are justice issues involved, even though ordination is not a human right. What is at stake for women in the use of a rigid opposition between a human right and a call by church authority is the exclusion of women *a priori* by reason of their sex from even being considered as persons possibly called by God to serve the church in ordained ministry.

13 *Constitution on the Church*, no. 1.

pally appointed parish administrators[14] since this practice places them in a realm traditionally held exclusively by ordained ministers. Existing for some time in certain parts of Central and South America, it is a relatively new occurrence in Canada and the United States. It has its roots in the teaching of Vatican II, the 1983 revised Code of Canon Law, and pastoral need. The Council proclaims that all members of the church, through their baptism and confirmation, are appointed by Christ and anointed by the Holy Spirit to build up the church in a continuing way. But further, quoting Pius XII, it teaches that the laity, in certain places and circumstances, can be entrusted with some ecclesiastical offices in order to insure the presence of the church's sanctifying activities.[15] The 1983 Code of Canon Law refers to the laity's ability to supply ecclesiastical ministry in canons 229–231, which describe their rights and obligations, and in canon 517, para. 2, which deals with the pastoral care of parishes where there is a shortage of priests.[16]

While some deny that the special circumstance of the shortage of priests, which justifies this pastoral practice, exists in North America,

14 There is not yet a uniform term for this new ministry. It is identified variantly as parish administrator, pastoral administrator, pastoral associate, pastoral worker, parish/pastoral coordinator, and pastoral life director. For a discussion of some canonical issues related to the title of this ministry, see Robert J. DeLand, *Some Implications of the Implementation of Canon 517, para.2 in the United States* (Leuven: Katholieke Universiteit Leuven, 1993) 38–39.

15 "Besides the apostolate which belongs to absolutely every Christian, the laity can be called in different ways to more immediate cooperation in the apostolate of the hierarchy, like those men and women who helped the apostle Paul in the Gospel, laboring much in the Lord (cf. Phil 4:3; Rom 16:3 ff.). They have, moreover, the capacity of being appointed by the hierarchy to some ecclesiastical offices with a view to a spiritual end." *Constitution on the Church*, no. 33. See also the *Decree on the Apostolate of Lay People*, no. 17.

16 Canon 517, para. 2 states: "If, because of a shortage of priests, the diocesan Bishop has judged that a deacon, or some other person who is not a priest, or a community of persons, should be entrusted with a share in the exercise of the pastoral care of a parish, he is to appoint some priest who, with the powers and faculties of a parish priest, will direct the pastoral care." *The Code of Canon Law in English Translation* (London: Collins, 1983).

many others acknowledge the significance of recent statistical studies[17] and expect the practice to increase, perhaps by as much as 20% in the coming years. In any case, it is a known fact that where such parish administrators do exist, women are in the majority of those so appointed and at the present time most of them are members of religious congregations.[18] A priest must be appointed the juridical head of the parish and, insofar as is possible, provide the sacramental ministries. Resident parish administrators, however, by episcopal appointment are in charge of the day to day life of the parish. Normally, they are charged with the Ministry of Worship, the Ministry of Education, the Ministry of Administration, and other pastoral responsibilities, such as fostering the leadership of the laity in the parish, pastoral care of the sick and those in need of pastoral counseling and assistance, and establishing diocesan awareness in the parish.[19] Their liturgical responsibilities usu-

17 Citing the 1990 *Official Catholic Directory*, DeLand, Some Implications, p. 9, reports: "Statistics in the United States show that in the last twenty years the number of priests has decreased by almost 7,000. An additional 6,000 more priests are retired or otherwise not available for ministry in parishes. The number of seminarians has decreased by 75%. The number of Catholics in the United States has increased by over 9 million. In 1990, 1,731 parishes representing 9% of the parish communities in the U. S. were listed as parishes without resident pastors." For a detailed statistical study on the dramatic changes in clergy supply, see Richard A. Schoenherr and Lawrence A. Young, *Full Pews and Empty Altars: Demographics of the Priest Shortage in United States Catholic Dioceses* (Madison, WI: University of Wisconsin Press, 1993).

18 Ruth A. Wallace, *They Call Her Pastor: A New Role for Catholic Women* (Albany, NY: State University of New York Press, 1992), 13; Marie Bergeron, *L'émergence des animatrices de paroisse dans le diocès d'Amos* (Québec: Université Laval, 1984). In a study I am conducting currently in selected eastern Canadian dioceses, nineteen out of twenty persons interviewed were women and eighteen of the nineteen women were members of religious congregations.

19 See the sample Pastoral Administrator Job Description and the Statement of Rights/Obligations of Pastoral Administrators in De Land, *Some Implications*, pp. 121–127. See also, Archbishop Thomas Murphy of Seattle, "Pastoral Care of Parish Communities," *Origins* 24 (1994–95) 755–764, one of many such statements on this position being published by particular U.S. bishops.

ally include the preparation for baptism, confirmation, marriage, reconciliation, the anointing of the sick and dying, and all arrangements for the celebration of the Eucharist, whenever it is possible. In many dioceses, according to circumstances, they preach at communion services or liturgies of the Word and they conduct funeral services.

Studies show that in most situations, the current generation of pastoral administrators are persons already known by priests and the bishop.[20] They are recognized as spiritual and pastoral leaders. Most of them have had previous professional experience in church or societal institutions as teachers, health care personnel, social workers, or pastoral counselors. In addition to their previous professional training, they have also acquired theological and pastoral education that is equivalent to that of ordained ministers and they have worked for some time, before being appointed pastoral administrators, in parish-based ministry. Thus, they bring a wide range of gifts to their office as parish administrators: theological education and professional training for ministry, pastoral ministry experience, and evident spiritual practice and maturity.

In the North American context, this rich array of leadership gifts readily available to the church has most often resulted from the historical commitment of religious congregations of women to equip their members for professional ministry, including theological and spiritual formation. Currently, women and men who are not members of religious congregations are also seeking these levels of professional and spiritual training for ministry. They are acquiring M.Div. and D.Min. degrees or their equivalents and more experience and training in the practice of spiritual disciplines. Most often, these members of the faithful pursue theological and pastoral studies without much institutional support. In some theological schools, such as Notre Dame University, there is a high level of support through grants, work-study programs, and scholarships. In most others, however, this has not yet been achieved. In the Canadian context, several religious congregations of women have been able to provide some support through scholarships for women, and some theological schools and pastoral training institutes have offered modest levels of support. What is still too rarely in evidence is parish or diocesan based financial and moral support of the faithful called to professional ecclesiastical ministry.

[20] See note 18.

Profiles of people currently appointed to this ministry of parish administration provide significant information with respect to the qualifications for this church position. Previous pastoral ministry experience, the M.Div. or an equivalent academic degree, and proven gifts for spiritual, pastoral, and organizational leadership are required.[21] While the situation is changing as noted above, it is members of women's religious congregations who most often have these qualifications already at hand. Further, if they are permitted by their respective congregational leaders to accept episcopal appointment to this ecclesiastical position, these women can commit to it on a full time basis and for an extended period of time. They are also often in a position to work for low levels of financial compensation.[22] For many of these reasons, they are often preferred to men from the already existing ministry of ordained deacons. There is a paradox here, if not a contradiction, since deacons hold permanent official public ministry status in the church, including the authority to preach, while at the same time they are usually far less trained and often not able to provide full time service to the church.

Studies also reveal that most parish members still prefer to have a priest as their resident pastor and some find it very difficult to accept a woman as their parish administrator.[23] At the same time, there is evidence that increasingly the service of women as full time parish administrators is being accepted. Parishioners appreciate in particular the collegial way in which their parish administrator works through parish councils, liturgical practice, and pastoral care. Some also believe that women should be ordained so as to be able to link their services in catechesis and pastoral care with the fullness in sacramental celebration.[24]

21 Murphy, "Pastoral Care of Parish Communities," 760.

22 My study of dioceses in Eastern Canada, for instance, has revealed that many of these women have retired in their early fifties with full pensions paid to their religious congregations from government supported school systems or health care institutions.

23 Wallace, *They Call Her Pastor*, 151–165.

24 Anecdotal reporting in my study, for instance, included such expressions as, "Our priest is a woman." In another situation, a terminally ill woman said to the parish administrator during a pastoral visit: "I wish you could say my funeral Mass when I die because you have cared for my spiritual needs." See also, Wallace, *They Call Her Pastor*, 47–85.

As to the women administrators' attitudes toward their ministry, all the women that I have interviewed manifested a deep sense of ecclesial identity rooted in a strong Christological faith and spirituality. While not without some difficulties with respect to their relationship to juridical authority, for the most part they experience themselves and their service in a positive light. They are committed to working within the present structure in creative and collaborative ways. They are dedicated to developing small and genuine communities of faith in places that otherwise would be neglected or abandoned. They believe they are building up the local church through this concrete experience of parish based ministry and, in the long run, also contributing to the universal church regarding the evolution of women as parish leaders. Their attitudes toward themselves as ordained are mixed. Several stated they would present themselves as candidates if it were available to them. Others said they probably would not, either because they find the present structure of ordained ministry too restrictive or because they feel no constraints on the exercise of their gifts in the current structure of delegated authority, or because they would find the permanency of ordination in conflict with their prior commitment to their religious congregation. In either case, all of them thought that women should not be excluded from ordination. Based on their pastoral experience and theological reflection, they know that women can image God and mediate the salvific presence of Christ effectively in parish ministry.

THEOLOGICAL REFLECTION ON THE CURRENT SITUATION

A rich pattern emerges from examining the situation of women in ministry today. Some express their evangelical commitment in society in forms of service that have no public connection with the church. Others respond to societal needs with ecclesial identity, that is, they are known explicitly in their workplace as Christians. Others serve in church institutions as volunteers or as professionals in already established roles as catechists, teachers, pastoral counselors, health care and social workers, and administrators. Others break new ground as chaplains, spiritual directors, theologians, canon lawyers, and parish and diocesan administrators.

A first obvious but not insignificant theological insight is the recognition of this marvelous plurality of gospel-inspired service and the celebration of its magnificence as a contemporary outpouring of the Holy Spirit. As a church, we may not yet be skilled enough to appreciate and validate genuine diversity without slipping into competitive and comparative estimations of the various forms of ministry. What we need is a transformational attitude that refuses to privilege one form over another, whether professional over volunteer or ordained over all other forms. To achieve such an attitude and to implement it structurally with justice, however, requires a radical transformation on the part of all members of the church, not the least of whom are those who are ordained since that form of ministry has been privileged historically for a very long time.

When we focus on women in ministry, understood more narrowly as official public service conducted in the name of the church and carried out with some form of church authority, some new challenges emerge for theological reflection and pastoral practice. One is the nature of delegated authority in the church when it is analyzed from the point of view of gender. Feminist theory points out that when authority is held exclusively by men, there is an inherent inequality in its exercise. Thus, even when women serve with the best of intentions, there are limits placed on the exercise of their gifts by reason of the fact that they labor in a structure of inequality. There are at least two paths to follow in redressing this situation. One is to ordain women and thereby include them in all levels of ministry by their investment with juridical authority. This position entails the belief that the existing hierarchical structure of church ministry is not inherently limiting, but only so in its current form. The exercise of authority by women, it is believed, would transform the pattern by endowing it with a fuller expression of the gifts given to the church through its members who are women as well as men. The church's proclamations about the collaborative and collegial character of ministry, this position holds, rings hollow without the inclusion of women as ordained ministers.

The other approach is to separate jurisdiction from ordination. Difficult as it might be to imagine unseating the centuries' old, doctrinally inscribed hierarchical character of the church, this stance does just that. It maintains that what is needed is an entirely new way of ordering ministry in which the differences among the gifts and services are related to each other structurally through the gospel-inspired proclamation of the discipleship of equals. Exactly how this vision would be incarnated organizationally has yet to appear, but it is certain to contain processes of genuine dialogue and consultation with some form of deliberative power given to members of the church as a discipleship of equals.[25]

A second issue arises when we look at the current situation of episcopally delegated "supply" ministry in which women play such a large part. Presently, official church teaching insists that all such ministry is temporary, justified only by a lack of priests. But two issues arise here. First, what is the duration of *temporary*? Some regions of the church have experienced the gifts of women, and lay men, as parish administrators for decades. When does the *temporary* become the norm? Second, what does it mean theologically to acknowledge that the women's gifts for pastoral leadership are authentic, genuine, and necessary *today* but not so *tomorrow*, if a priest becomes available? This, it seems to me, is to give precedence to the juridical over the proven pastoral practice of women. What is needed instead is the honest recognition of the gifts of women for ministry in the church as valid in themselves, and not just on a temporary basis.

A third issue concerns the place of pastoral practice in effecting change in church structure and teaching for the good of the believing community. In too many instances, present pastoral practice is generating a separation between the word and sacrament, between pastoral care and symbolic fullness in liturgical celebration. The teaching of Vatican II on the unity in sacramental priesthood between cult and servant parish leadership is being severed unnecessarily and with grave doctrinal and pastoral harm. Weekly celebration of the Eucharist may not have to be the norm, but its absence for long periods of time as the unifying sacramental celebration of a faith community is certainly contrary to the core of Roman Catholic identity.

Is it not time to break the tradition of the exclusion of women from ordained ministry? Feminist theology, for all practical purposes, has critiqued this tradition, based as it is on outdated subordination theories regarding women's identity and therefore fittingness for sacramental

25 Ruether, *Sexism and God Talk*, 193–213; Schüssler Fiorenza, *Discipleship of Equals*, 290–372; Eugene C. Bianchi and Rosemary Radford Ruether, eds., *A Democratic Catholic Church: The Reconstruction of Roman Catholicism* (New York: Crossroad, 1992).

signification. Women's pastoral practice, already decades old in some parts of the church, has established the efficacy of their gifts for linking the word and pastoral care ministries with sacramental sign. A gradualist approach to this change would at least permit their admittance to the order of deacons, since there is already historical precedence for this practice.[26] While the ordination of women to the diaconate would not solve all the problems in the church's current pastoral practice, it would certainly eliminate many of them and finally offer some sign that the church means what it says without reservation when it declares that women's gifts are needed and welcomed in ecclesiastical ministry.

The obstacles are many to the removal of all conditions on the exercise of ministry by women in the church. Current church policy, for instance, proposes that the shortage of priests in certain parts of the world will be met by moving priests around internationally since there is an increased number of clergy available in certain regions of the world.[27] But one could ask why such a policy has not been effective thus far in Latin and Central America, where the shortage of priests has existed already for decades, and why it would be needed in countries like the United States and Canada, where there is already an abundance of ministers qualified in every way for full pastoral leadership except by gender or a commitment to lifetime celibacy. In addition to the cross-cultural issues involved, why should other countries, where the church is in early stages of development or of education to the ecclesial vision of Vatican II, be drained of their resources to serve in countries with plenty of resources? Further, what understanding of sacrament and of priesthood in terms of its relationship to the community is at play in this strategy?

But if the obstacles are great, so is the commitment of many to the full inclusion of women in church ministry. If the last thirty years have brought such enormous change in the theology and practice of church ministry, especially as women claim the full extent of their voice and giftedness for ministry, then what will the next thirty years show as

26 Canon Law Society of America Ad Hoc Committee, "Canonical Implications: Ordaining Women to the Permanent Diaconate," *Origins* 25 (1995–96) 344–351.

27 Archbishop Pio Laghi, "Report on the Redistribution of Priests," *Origins* 20 (1990–1991) 681–685.

the Holy Spirit continues to help the church reach its full stature as the sacrament of God's presence among us? What is clear is that women will continue to show persistent courage in the creative reconstruction of church life and ministry.

Women *at* Eucharist *as* Crucible *and* Hope

Theresa F. Koernke, IHM

INTRODUCTION[1]

Before the reforms of the Second Vatican Council (1962–1965), most Catholics understood the word "Eucharist" to refer to the sacrament reserved in the tabernacle, exposed for adoration, or received at Mass, and the so-called "words of consecration" as sole moment of significance. This intense emphasis on those words of the priest in scholastic theology and its associated catechesis led to theological neglect of the meaning of the proclamation of the Scriptures, the significance of the entire Eucharistic Prayer (the Canon), and to the role of the other members of the Assembly. Although the popes occasionally discouraged ministerial priests from celebrating without anyone else present, the practice of "private Mass" endured.[2] That classical world view and its correlative theology viewed ministerial priests as the "full Church," each as an *Alter Christus* acting in the person of Christ now seated at the right hand of God, each having the personal power, as Jean Baptiste Lacordaire extolled, "to command the Almighty to leap down upon our altars."[3] Only with the confluence of the biblical and liturgical movements, each with its own stages of development, would these emphases on the actions of the priest and on the "moment of consecration" come under the scrutiny of the riches of the Catholic Christian Tradition.[4]

1 The contents of this reflection have been shared with several communities of women religious throughout the American Catholic Church. It is fitting that they be a gift to Margaret R. Brennan, I.H.M., who has encouraged us to be goads to the conscience of the Church, and given me the passion to speak the Truth with affection.

2 The current Sacramentary issued by Pope Paul VI in 1970 still contains the performative oxymoron, Missa sine Populo (Mass without People), a vestige of this scholastic theological synthesis.

3 This reference frequently appears on ordination announcements. Its origin is unknown to this writer.

4 For a succinct account of the stages of development of the Liturgical Movement, see Gilbert Ostdiek, "Ritual and Transformation: Reflections

Others may only know the celebration of the Eucharist since Vatican II. During these years, the New Testament foundations of the full celebration of the Eucharist as an action of all the priestly people struggle for acknowledgment. These two broad experience bases explain the tremendous variety in the attitudes, questions and concerns surrounding the Eucharist. Further, the rise of feminist consciousness, as well as the awareness that scholastic theology no longer carries the weight of the articulation of the faith of the Church in a postmodern world, lends added poignancy to the description of the current situation of Catholic Christian women as standing in a crucible: a place in which apparently and really disparate concerns are melted together under intense heat. Some of these concerns are: the experience of male-domination in Church life; sex-exclusive language referring to God and persons; the meaning of the many modes of "the real presence of Christ" at Eucharist and otherwise; the relation between Eucharist and the moral life; the value of celebrations of communion services *in lieu* of the full celebration of the Eucharist; the significance of daily Mass; healthy diversity in relation to subjectivism and relativism; the call of competent women to the colleges of presbyter and bishop (ministerial priesthood).[5] Indeed, the issue of the celebration of the Eucharist has become a crucible in which women stand or, in some cases, from which they flee as beyond hope of reform.

To be sure, questions and concerns are clear indications of deep, often unspoken, controlling assumptions about the meaning of being baptized into the One Priesthood of Christ in this holy and sinful Church. And *a fortiori* they are clues to operative understanding of the Eucharist and of participation in it with integrity. For example, women are concerned about why some persons either do not celebrate the Eucharist everyday, but only on Sundays or not at all, or about gatherings called "eucharist" at which only women are present. While many women find no problem with a male-only college of presbyters, others experience deep pain and anger at this exclusion. Many women are high-

on the Liturgy and the Social Sciences," *Liturgical Ministry* 2(Spring 1993): 38–48.

5 See Inter Insignores (On the Question of the Admission of Women to the Ministerial Priesthood, 15 October 1976) *Acta Apostolicae Sedis* 69(1977): 98–116, Sacerdotalis Ordinatio (On Priestly Ordination, 22 May 1994) for the most recent statements of the Magisterium.

ly sensitive to the fact that all language referring to God (including the language used by Jesus) is used by way of analogy from human experience, and that no language can fully capture the reality of the living God. And, in justifiable reaction to the power of androcentric language to the full humanity of women, some women refuse to ever use male metaphors for God, even though they are metaphors from the living Tradition.[6] Still others imagine that female metaphors for God are heretical. While some women presume that the only "real presence of Christ" is obtained in the sacrament, others have come to appreciate the real presence of Christ by the power of the Spirit at all times, have come to celebrate the Eucharist on Sundays only, have wondered why we need to celebrate the Eucharist at all, or have questioned why any gathering that recalls the Last Supper may not be called "eucharist."

It would be an understatement to say that the celebration of the Eucharist is the public, social activity of the Church in which every major issue in Catholic Christian life is manifest and wrestled. Indeed, the arrangement of chairs is architectural ecclesiology, as is the positioning of the baptismal font in relation to the ambo, and the location of the tabernacle.[7] And certainly, what anyone thinks about the nature of the Church is manifest in various responses to the issue of women as acolytes or in any other ministry, or whether any lay person ought to preach in the context of the Eucharist.[8]

To advance reflection *as Catholic Christian women* on the meaning of the celebration of the Eucharist, and to enable appreciation of the origin of assumptions about the Church and the Eucharist that ground concerns and practices, the following considers (1) the impact of "historical

6 For an explication of the meaning and function of "metaphor," see Sandra Schneiders, I.H.M., *The Revelatory Text: Interpreting the New Testament as Sacred Scripture* (New York: HarperCollins Publishers, 1991), 28–36. For extended consideration of language in public worship, see Gail Ramshaw, *Searching for Language* (Washington, D.C.: The Pastoral Press, 1988).

7 See Bishops Committee on the Liturgy, National Conference of Catholic Bishops, *Environment and Art in Catholic Worship* (Washington, D.C.: United States Catholic Conference, 1978), 78–80.

8 Mary Collins, O.S.B., "Women in the Liturgical Assembly," chap. in *Contemplative Participation: Sacrosanctum Concilium Twenty-five Years Later* (Collegeville, MN: The Liturgical Press, 1990), 23–35.

consciousness" upon the celebration of the Eucharist, (2) the process of theological reflection, (3) a participant observation of liturgical practice among women religious, and (4) an elucidation of the depth structure of the Mass.[9]

HISTORICAL CONSCIOUSNESS AND THE EUCHARIST

Human thinking is increasingly marked by ever expanding "historical consciousness." Theologians speak of paradigm shifts—changes in the way humans make sense of reality—away from the classical world view inherited from Plato and Aristotle and espoused by Saint Thomas and others.[10] Clearly, the contributions of Copernicus, Galileo and Newton challenged the prevailing view of reality and therefore of the relations of God to creation.[11] From the mid-eighteenth century on, the modern scientific world view seriously questioned the adequacy of the classical or scholastic way of talking about reality—God, creation, redemption, Christ, Church, liturgy, the full humanity of women. For example, this period notes the use of the historical-critical study of scriptures which, among other things, challenges a naive biblicist reading for Church organization and stable leadership, as well as of the accounts of institution. Further, the modern scientific world view encouraged an explosion of confidence in what the human mind can achieve and an almost unbridled attention to the "subject" as sole norm for determining reality. From the turn of this century, philosophers, the-

9 This terminology is taken from the hermeneutical theory of texts of Paul Ricoeur and applied to liturgical ritual activity by Joyce Ann Zimmerman, C.Pp.S., *Liturgy as Living Faith: A Liturgical Spirituality* (Scranton: University of Scranton Press, 1993), 36–51. She notes that the depth structure of a text is the dynamic that makes it what it is. In the case of the Mass, the depth structure is constituted by the soteriological moment in the Liturgy of the Word, and the eschatological moment in the Eucharistic Prayer and communion.

10 Robert M. Gascoigne, "The New Cosmology: Science Ponders Divine Creation," *The Autralasian Catholic Record* 71(July 1994): 330–340.

11 N. Max Wildiers, "From Copernicus To Darwin" and "Crisis in Religious Thought," chaps. in *The Theologian and His [sic] Universe: Theology and Cosmology from the Middle Ages to the Present* (New York: The Seabury Press, 1982), 83–129.

ologians and scientists have realized that, as marvelous as it is in some respects, the scientific view of the world is itself inadequate to the human effort to comprehensively make sense of reality.[12] Indeed, these postmodern sensibilities have served to challenge modern scientific assumptions about human activity that has led to disregard and destruction of our planet for personal gain. On the other hand, if postmodern historical consciousness is marked by the realization that scholastic theology and its assumptions no longer carry the weight of articulating the faith of the Church, then neither can the modern world view, with its assumptions about the power of reason and the disregard of the interconnectedness of all creation, worthy of our complete trust in making an account of the hope that is in us (I Peter 3:15–15).

Reflection on the tensions surrounding the Eucharist from postmodern historical consciousness implies not only the fact that persons have acted and events have occurred from within a given cosmology or philosophical world view, but also that persons are thinking and acting today from out of one or a combination of cosmologies and their corresponding views of God, Christ, Church, leadership, worship, and the call of women to ordained ministry. It is understandable that these insights would influence the central act of worship of God we call the Eucharist.

On Theological Reflection

Whenever an apparently new issue arises in the life of the Church, a trialogue serves to focus the issue: (1) What is the contemporary faith experience that produces the issue? (2) What, if anything, have former ecclesial practice and theology to offer? (3) What new insights may be gleaned when both past and current experience are submitted to the best means of interpretation available? Indeed, in the light of current interpretation from the stance of faith, clarification of issues has and continues to demand a critique of inadequate responses in the past or present, whether of our own, those of official sources, or of theologians.[13] *Ecclesia semper reformanda.* In short, historically conscious faith

12 For an early theological attempt to appropriate the negative impact of the modern world view, see Romano Guardini, *The End of the Modern World: A Search for Orientation*, trans. Joseph Theman and Herbert Burke, ed. Frederick D. Wilhelmsen (New York: Sheed & Ward, 1956).

13 Sandra M. Schneiders, I.H.M. engages in a similar discussion in a video

that seeks understanding knows of healthy relativity, both for previous practice and theology, and for current subjective experience.

A PARTICIPANT OBSERVATION

Even a purely visual note of the celebration of the Eucharist in gatherings of women religious reveals an improvisational attitude, indeed, a degree of divergence from or layering onto the liturgical order that often endangers the objective meaning of the Eucharist for Catholic Christian life. Are there sources of this phenomenon?

In the years after Vatican II, efforts were made to overcome the rubricism attributed to the Tridentine liturgical reforms. The Council of Trent (1545–1563) responded to many abuses. For example: wagering and rooster fights at Mass; extended elevations of host and cup; hundreds of cross gestures during the Canon; lengthy genuflections at the words of institution; and, placing of cadavers on the altar (an admittedly erroneous attempt to communicate the fact that the Mass is the self-offering of Christ). Accounts of the period of time before the Council of Trent reveal that ministerial priests were straying from or adding things into the context of the Mass *to the point that their actions inhibited understanding of the true meaning of the Mass*.

In response to these abuses, the Council of Trent mandated the reform of the Mass, promulgated in the 1570 Roman Missal of Pope Pius V. But these reforms were engaged from within the classical world view, the pyramidal diagram of which appeared in our catechisms and Benziger Church History books. Therein, directives were printed in red (hence, the term *rubrics*) to direct the priest, the central figure in the classical theological account of the Eucharist, in doing and saying what the rite was thought to mean in that world view. The purpose of the legislation was to prohibit the abuse of behavior that gave false meaning to the rites.[14]

---

presentation for the Theological Education Project, Cycle I (Monroe, MI: Sisters, Servants of the Immaculate Heart of Mary, 1995).

14 See the *Canons and Decrees of the Council of Trent*, Original Text with English Translation by H.J. Schroeder (London: B. Herder Book Co., 1941), Sessions Twenty-two, "On the Mass," and Twenty-three, "On the Sacrament of Orders."

So long as priests adhered to the meaning of the depth structure of the Mass (see Note 9) according to the faith of the Church, then the following of the directives printed in red did not make for "rubricism," but for doing what the Church intends and understands to be going on. But, on the whole, neither clergy nor laity were well catechized or well educated in the theology of the Mass, and a sense grew that the Mass was effective of God's grace simply because the externals were observed. Such was and is rubricism: the attitude that following the rubrics, i.e., the mere external actions, constitutes validity. *It was and is this severe attention to the role of the priest and to the rubrics, rather than to the depth structure of the Mass, which would lead to a sense of near magic surrounding the liturgy.*[15] In short, the purpose of Trent's reforms was not to say that the rubrics make a sacrament a sacrament, but to keep the clergy from diverging from or layering onto the rite by behavior that violated the depth structure of the Mass according to the faith of the Church.

In our efforts to understand current attitudes and reactions on all sides of issues surrounding the Eucharist, it is significant to note that the entire scholastic Tridentine teaching only accounted for the actions of ministerial priests in relation to Christ in the celebration of the Mass. This theological, catechetical and pastoral emphasis had the infelicitous consequence of describing the faithful as passive recipients of grace, of which the ordained minister was the channel, so long as serious sin was absent.[16]

The insights of the nineteenth century biblical movement and the initial stage of the liturgical movement, which drew on modern textual research, would gradually question the adequacy of the classical scholastic account of the Mass. In this century, Pope Pius XII issued three major documents: the 1943 encyclical *Divino Afflante Spiritu*, which acknowledged the good of the historical critical method of textual study; the 1945 encyclical *Mystici Corporis*, which continued the gradual retrieval of communal ecclesiology; and the 1947 encyclical *Mediator Dei*, the first

15 For a brief description of the reductionistic understanding of "matter and form," see Herbert Vorgrimler, *Sacramental Theology*, trans. Linda M. Maloney (Collegeville, MN: The Liturgical Press, 1992), 160–163.

16 For a summary and critique of this theology, see Edward J. Kilmartin, S.J., *Church, Eucharist and Priesthood: A Theological Commentary on The Mystery and Worship of the Most Holy Eucharist (Dominicae Cenae*, Holy Thursday Letter of Pope John Paul II, 1980 (New York: Paulist Press, 1981), 18–22.

mandate for the restoration of the public worship of the Church. More recently, the documents of Vatican II voiced the profound retrieval of the conviction of faith that all the baptized share in the Spirit of the One Priesthood of Christ. This retrieval does not deny the essential need for ordained leadership in the Church, but does assert our consolation: Christ among us by the power of the Spirit to the glory of God, always and everywhere. Specifically, the *Constitution on the Sacred Liturgy* (CSL) renewed attention to the *sign value* of our common worship, that is, to what and how liturgical words, gestures, signs, and symbols *mean* (CSL, 7).[17] In addition to criticism of minimalistic use of symbols such as water, bread, wine and oil, this attention to *signification* has served to highlight the symbolic function of leadership of the Eucharist *in the person of Christ*, and to question the present exclusion of women from presiding over the Eucharist.[18]

In the face of this core retrieval, a question arises: If it is true that we all share in the Spirit of the One Priesthood of Christ, how are the signs, symbols and patterns of behavior within public worship, especially the Eucharist, to express this meaning? Clearly, the use of the vernacular, and the extension of the ministries of lectoring and of bread and cup, are happy expressions of this radical insight. However, in reaction to centuries-long experience of the Mass as the domain of the priest, and of the laity as relatively passive "hearers" or observers of ministerial priestly actions, these years also witness to a variety of attempts, having little to do with the depth structure of the Mass, to engage persons in participation. From the deep desire to assure persons that God in Christ is not distant, our places of worship have frequently taken on the aura of cozy living rooms or busy marketplaces. Being casual and colloquialhas

17 See also, Edward J. Kilmartin, S.J., "Human Sciences and Sacramental Theology: Symbolic Perception," chap. in *Christian Liturgy: Theology and Practice*. I. Systematic Theology of Liturgy (Kansas City, MO: Sheed & Ward, 1988), 26–29.

18 A psychoanalytic analysis is presented by Mary Ann Jordan, "Liturgical Presidency: Symbol in Crisis," *Assembly* 17 (November 1990): 502–503. For a post-scholastic theological presentation of the significance of the term *in persona Christi*, see Edward J. Kilmartin, S.J., "Bishop and Presbyter as Representatives of the Church and Christ," chap. in *Women Priests: A Catholic Commentary on the Vatican Declaration*, eds. Leonard Swidler and Arlene Swidler (New York: Paulist Press, 1977), 295–302.

had the consequence of creating another experience of diverging from or adding things onto the Mass that obscure its depth meaning.

Since for centuries the term "priesthood" referred solely to ministerial priests, it is occasionally assumed that the retrieval of the truth of the Universal Priesthood of All Believers now implies that anyone may now do what the ordained minister is deputed to do. Several issues were and are in play around this assumption. The Letter to the Hebrews uses the metaphor, High Priest,[19] to speak of Jesus as Mediator between God and humanity, and yet, the author does not intend to say that Jesus is a ministerial priest. Rather, the author's purpose is to distinguish the "once-for-all" saving activity of Jesus from the need to repeat animal sacrifices according to Jewish Law. That is, if Jewish priests repeated the offering of animals in the past, so now, *by contrast*, Jesus offers humanity to God by offering his own person, in and through the flesh he shares with humanity. Because Jesus is the Eternal Word/Wisdom of God made flesh, his self-offering (sacrifice) is in no need of repetition.[20]

The subtleties of this biblical truth were all but lost in classical scholastic theology. And when that theology identified ministerial priests with the risen Jesus, the expression "You are a priest forever according to the order of Melchizedek (i.e., by using bread and wine)"[21] was thereby attributed to ministerial priests. The consequences of this identification (one could say "mystification") in terms of clericalism are well documented. Therefore it is understandable, if regrettable, that persons have latched onto the notion of the Universal Priesthood of All Believers, and have understood it as referring to ministerial priesthood

19 See Hebrews 5:6; 5:10; 6:20; 7:1; 7:10; 7:11; 7:15 and 7:17 for reference to the analogy and distinction between the priesthood of Melchizedek and that of Jesus.

20 See Myles M. Bourke, "The Epistle to the Hebrews," in *The New Jerome Biblical Commentary*, eds. Raymond E. Brown, S.S., Joseph A. Fitzmeyer, S.J., and Roland E. Murphy, O.Carm. (Englewood Cliffs, NJ: Prentice Hall, 1990), 920–941.

21 In Genesis 14:18, Melchizedek, king of Salem, blessed God over bread and wine, hence the references in Hebrews to Jesus whose "priesthood" we know through his transformation of the meaning of blessing God over bread and wine.

for all.[22] The challenge to genuinely appropriate the truth of the Universal Priesthood of All Believers remains.

For many persons, the 4 December 1963 promulgation of the *Constitution on the Sacred Liturgy* seemed to encourage the setting aside of the directives printed in red in our liturgical books. Divergence or layering onto the Mass took the form of readings other than scripture and Christmas pageants in place of the Liturgy of the Word; presentation of cleats, pompoms and other things at the Preparation of the Gifts; and dramatization of the words of institution or choral reading of the Eucharistic Prayer, and so on. With the good intention to appropriate the truth that all the baptized share in the Spirit of the One Priesthood of Christ, many persons reasoned that these divergences from and layerings onto the rite were expressions of genuine regard for the diversity of cultures and of a movement beyond rubricism toward "full, conscious and active participation" (CSL, 14).

As participant observers of liturgical celebrations, women might ask: How free are we of the scholastic theology we imagine ourselves to be outgrowing? Given the lingering vestiges of scholastic theology, have many women equaled participation in the liturgy with the role of the ministerial priest? Are the words of instiution all that really matter? Has the centuries-long sole attention to the role of the priest falsely convinced women that membership in the full Church means membership in the college of presbyters? Does inculturation with feminist sensibilities imply disregard for the depth structure of the Mass?[23]

The integration of the liturgical theology of Vatican II and beyond with the data of contemporary ritual studies and hermeneutics has led to a critical reassessment of our initial efforts at renewal. The project asserts that if the chief rites of the Church are efficacious, that is, authoritative, precisely by meaning or signifying the source of their power (the life, ministry, death and resurrection of Christ), *then doing anything that clouds the meaning of the rites for the participants, clergy or*

22 This narrow conception of the meaning of the "blessing" of bread and cup seems to be espoused by Rosemary Radford Ruether, *Women-Church: Theology and Practice of Feminist Liturgical Communities* (San Francisco: Harper & Row Publishers, 1986), 90–91.

23 Theresa F. Koernke, I.H.M., "The Roman Liturgy and Inculturation: Have We Accommodated What We Should Have Inculturated?" *Pastoral Music* 19 (June–July 1995): 30–34.

laity, has the potential to be a lie in motion. Such behavior has the potential to stunt the very truth on which we stake our hopes for the liberation of women. And these observations hold true for those who would cling to the liturgy of Trent.

Today, and under the heavy cloud of *Sacerdotalis Ordinatio* (22 May 1994) and the subsequent pronouncement of the Congregation of the Doctrine of the Faith (28 October 1995), women face several challenges: to so know the meaning of being baptized into Christ that, while working to dissolve the scholastic theological blocks to the call of women to the ministerial priesthood, they do not confuse the current refusal with expulsion from the Body of Christ; to so know themselves as to distinguish the deep psychological pain over the current refusal from the fact that patriarchy is not greater than the fire of the Spirit; to so know the history of the development of practice and theology of the Eucharist that there is no temptation to confuse the liturgy of Trent with what Jesus gave the Church; and to know the depth structure of the Mass as source of the demise of patriarchy.

Contemporary ritual studies enable the understanding that theological reflection on all the behaviors and words of the Mass presupposes crucial "givens": specific kinds of religious ritual are repetitive, interpersonal, and value-laden behaviors of a social group; they are intended to be an expression of its experience of the Holy One; the chief rites of a group (here, Initiation and the Eucharist) are the source of social identity and self-understanding; this religious activity is itself *governed by some previously established norm* (depth structure); and finally, the chief religious rites of a group have socio-ethical implications.[24]

These data show that the rites of a people are vitally important acts, not simply of psychological memory, but of *anamnesis*, a kind of remembering that renders the force of the past event present.[25] In the holy and sinful Church, *anamnesis* is the calling to mind, that is, the acknowledging in the present, that the Event (the self-offering of Jesus on the Cross) that brought the Church into existence is somehow present, continues to exert influence, and orients the Church toward the future. *Therefore,*

[24] Theresa F. Koernke, I.H.M., "Towards an Ethics of Liturgical Behavior" *Worship* 66 (January 1992): 25–38.

[25] Brevard S. Childs, *Memory and Tradition in Israel*. Studies in Biblical Theology, 37. (London: S.C.M. Press, Ltd., 1962); Jean-Claude Sagn, "La mmoire du coeur" *Vie Spirituelle* 132 (1978): 184–199.

these rites or "liturgical orders" operate on the principle that their meanings have been established (encoded) by someone other than the participants. Given the many cultures in which the Church prays, this principle does not deny inculturation. Rather, it implies that there are certain consistent coordinates without which the Eucharist is not. Indeed, because of the social nature of our knowledge of God, these rites, especially the Eucharist, are not simple alternatives for doing or saying something.

Religious life is constituted by charisms from the Spirit to live by the public commitment to simplicity of life, love of others through celibate chastity, and discernment of God's will in relationship with a specific community. At crucial junctures in its history, religious life flourishes by retrieving its role to goad the conscience of the Church. And it does so, not because its members are without sin, but because they have grown wise and fearless precisely by naming it, and by reminding the Church that the free exchange of ideas is inherent to the search for the Truth who is God. Reformist feminism, which is consistent with the charism of religious life, does not imply a complete rejection of the past, but rather a radical reappropriation of the Tradition, the gift of Christ by the power of the Spirit. The vibrant tradition of fearless wisdom, embodied by women like Catherine of Siena and Theresa of Avila, serves women well in a retrieval of the depth structure of the Mass in order to stand firmly in the Eucharist-crucible of our times.

THE DEPTH STRUCTURE OF THE MASS

Because the Church is the continuation of the bestowal of the Spirit of Christ in the Incarnation, the Church has always attempted to see what it does at liminal moments in human life in the light of the meaning of the life, death and resurrection of Christ. Liminal moments are those events, such as birth, death, illness, commitment, deputing leaders, and so on, that defy our full intellectual grasp, but which call for interpretation.[26] Thus, Initiation interprets Christian life as buried with Christ for the hope of resurrection; the Eucharist is the corporate way

[26] George Worgul, "Sociology-Anthropology: The Function of Ritual in Human Community," chap. in *From Magic to Metaphor: A Validation of the Christian Sacraments*, with a foreword by Piet Fransen, S.J. (New York: Paulist Press, 1980), 70–110.

Christians reaffirm their origin in the ministry and Cross of Jesus; Reconciliation celebrates corporate healing while Anointing is directed to interpretation of personal suffering; Marriage between woman and man delights in the human share in the ongoing creative power of God; and in the deputation of its stable leaders, the Church recalls that there is only One Head of the Church. *But any given theological explanation of these chief rites has always depended on the current understanding of how the risen Christ is related to human history, that is, on the understanding of Resurrection and Ascension.*[27]

With the collapse of the Roman Empire (476 C.E.) theology lost touch with the manner in which both the scriptures and the liturgical rites came into existence. The scriptures, read in a naive biblicist way, seemed to imply that the risen Jesus had ascended to "the right hand of God" in a physicalist manner, and thus was not present to the Church at all times. Yet, the liturgy asserted a real encounter with Christ. Given this problematic, systematic theology needed to explain the way in which Christ was present in the sacraments. An example of such an effort is the work of a sixth century author, one "Dionysius." Assuming the platonic heavenly hierarchy of being, he situated Christ "at the right hand of God in heaven" at the pinnacle of the spiritual beings. Since in this world view the spiritual world patterns the material world, "Dionysius" placed the pope and clergy at the pinnacle of the ranks of being in the Church. The association of the clergy and pope with Christ was immediate; it had the effect of equating ordained ministers with the "full Church" and of describing the faithful as "merely baptized." Logically, ordained ministers were interpreted as conduits of the grace of Christ in the celebration of the sacraments.[28] Most subsequent theologians presumed that the author was "Dionysius the Areopagite," companion of Paul at Athens (Acts 17:33–34), and that his description of the order and rites of the Church were, therefore, what Jesus had directly

[27] Gustave Martelet, "The Anthropology of the Resurrection," chap. in *The Risen Christ and the Eucharistic World*, with an introduction by Thomas Corbishley, S.J., trans. Rene Hague (London: Collins, 1976), 60–95.

[28] See Thomas L. Campbell, trans. and ed., Dionysius the Pseudo-Areopagite, *The Ecclesiastical Hierarchy* (Washington, D.C.: University Press of America, Inc., 1981), for an exposition of the text that would influence the work of Thomas Aquinas and subsequent theologians on the role of the ministerial priest.

determined. By the ninth century, theology of the Mass focused solely on the role of the ministerial priest, what happened to bread and wine, and how the Mass could be the sacrifice of Christ, in near oblivion to the radically social-ecclesial meaning of "Body of Christ." The tenth century witnessed this oblivion in the rise of "exposition-for-viewing" of the reserved sacrament, since regular communion had ceased.[29]

The twelfth and thirteenth centuries witnessed the increased use of the language of Aristotelian philosophy: for example, substance, accidents, efficacy and causality. In this theology, the ordained priest was said to receive the personal power to repeat and actualize the words of Jesus, and hence to be an instrumental cause of Christ's presence through the change of the substance (transubstantiation) of material elements. Well into this century, the Saint Andrew Missal explained that, at the words of institution, the priest consciously associates himself with Christ at the Last Supper.[30] In short, scholastic theology of the Mass no longer accounted for the various modes of the presence of Christ in virtue of baptism, but only for a narrow understanding of the presence of Christ in the sacramental bread and wine, to the neglect of the Liturgy of the Word, the entire Eucharistic Prayer, and Communion. Yet, even among those who hail the achievements of contemporary theology, the centuries-long influence of scholasticism upon subjective expectations about the meaning of the Mass continues to effect resistance against the spiritual-liturgical renewal encouraged by Vatican II.

The current challenge is to retrieve and appreciate the depth structure of the full celebration of the Eucharist, and thereby contextualize questions about the Universal Priesthood of All Believers in relation to ministerial priesthood and hoped-for call of competent women to that service of the Body of Christ. The following is a brief presentation of the *origin and depth structure* of the celebration of the Eucharist.

29 Nathan Mitchell, "Controversy and Estrangement," chap. in *Cult and Controversy: The Worship of the Eucharist Outside Mass*, Studies in the Reformed Rites of the Catholic Church, Volume IV (New York: Pueblo Publishing Company, 1982), 66–128. Prior to the custom of building large tabernacles, the sacramental bread was reserved in the sacristy in the same place with the Gospel Book and oils.

30 Dom Gaspar Lefebvre, O.S.B. and the Monks of the Abbey of St. Andrew, eds., *Saint Andrew Daily Missal With Vespers For Sundays and Feasts* (St. Paul, MN: The E. M. Lohmann Co., 1958), 823.

Clearly the disciples of Jesus did not fully understand who he was before his execution by crucifixion. Only afterward did they come to know the meaning of his life and actions, especially of what he had done at a ritual meal sometime close to the end of his life. Indeed, the four New Testament references to that ritual meal show a growing awareness of its meaning in relation to the ministry and Cross of Jesus.[31]

As good Jews, Jesus and the disciples had gone to the synagogue. And for a short period of time after the experience of the resurrection and before their expulsion from the synagogues, the disciples continued to observe Saturday, the Jewish Sabbath, for prayer and to hear the Jewish scriptures proclaimed. This synagogue Word Service is a source of the Liturgy of the Word.[32] They gathered on the first day of the week, Sunday, to renew their belonging to the Body of Christ and their commitment to the New Covenant in the Blood of Jesus. Before long, the disciples of Jesus stopped going to the synagogues, but gathered to read the Jewish scriptures before they ate the ritual meal, thus establishing the basic pattern of the full celebration of the Eucharist.[33]

Justin Martyr (✠156–166 C.E.) writes that the faithful gather on the Day of the Sun; the one who presides greets the Assembly in the name of the Lord Jesus; the scriptures are proclaimed from the Law, the Prophets and from the memoirs of the apostles (Letters and Gospels). Then, the one who presides proclaims the Great Prayer of the Blessing of God in the hearing of the Assembly, the prayer in which the Church reminds God of all the works of salvation, the greatest of which is the revelation in Jesus who taught the Church to pray in this way (Justin

31 Jerome Kodell, O.S.B., "The Accounts of the Last Supper," chap. in *The Eucharist in the New Testament*, Zacchaeus Studies: New Testament, ed. Mary Ann Getty (Collegeville, MN: The Liturgical Press, 1991), 53–67.

32 Eric Warner, "The Jewish Liturgy at the Time of Primitive Christianity," chap. in *The Sacred Bridge: The Interdependence of Liturgy and Music in Synagogue and Church during the First Millennium* (London: Dennis Dobson/ New York: Columbia University Press, 1959), 1–16.

33 Initially, the disciples read the Jewish scriptures, interpreted in the light of the experience of Jesus as the power of God to save. Gradually, the reading of the letters of Paul followed, and then the "gospel" literature that interpreted the meaning of the life of Jesus and the consequences of following him. *Whether or* to what extent the four New Testament gospels constitute a distinct literary genre is a debated issue among scholars.

Martyr, *First Apology*, 67).[34] This depth structure of the celebration of the Eucharist is patterned on the depth structure of the Christian relationship to God: by the breath of the Spirit, God speaks through Jesus the Word made flesh; through, with and in Christ, the presider praises and thanks God in the hearing of the Assembly, and the Church affirms its own reality—Christ and his Members —in Communion.

What did the last meal of Jesus with his disciples look like, especially the prayers?[35] Before a regular meal of bodily satisfaction, the leader of the group prayed grace before meals by taking bread and blessing God over it and by accounting all the great works of God for Israel, especially God's creation of the people and remaining present to them. The leader passed the bread *in silence*. Any good Jew would have understood that in eating this bread they were eating the history of the people whom God had created. This eating was understood as a real encounter with God, truly present, in and through the people, and therefore, an acceptance of the responsibilities of belonging to this corporate reality.

After the meal of bodily satisfaction, the leader prayed grace after meals, blessing God over a cup of wine for having made the Covenant at Sinai and for giving the privilege of keeping the Covenant through observance of the Law. Then the leader passed the cup *in silence*. Any good Jew would have understood that to drink from this cup implied reaffirmation of the Covenant and the assumption of ethical living. Having done this, the leader prayed that God would hear the petitions of Israel.

Sometime close to the end of his life, Jesus ate such a meal, but added something that his disciples understood only after his death. Jesus prayed grace before the meal by blessing God over bread, ate a portion, and gave it to his disciples. When he did so, he broke the usual silence with the words: "Take, eat, this [is] my body." When supper was over, Jesus prayed grace after meals, praising and thanking God for the

34 For the complete text, see Lucien Deiss, C.S.Sp., *Springtime of the Liturgy: Liturgical Texts of the First Four Centuries*, trans Matthew J. O'Connell (Collegeville, MN: The Liturgical Press, 1979), 93–94.

35 This synthesis is based on the scholarship of Jewish texts that would have reflected the pattern of ritual meals during the time of Jesus. For a thorough presentation, see Enrico Mazza, *The Eucharistic Prayers of the Roman Rite*, trans. Matthew J. O'Connell (New York: Pueblo Publishing Company, 1986), especially pp. 1–29.

Covenant and the Law, and drank from the cup. And when he did so, he broke the usual silence with the words: "This [is] the New Covenant in my blood."

For centuries, the eating of the bread had implied the radical encounter with God in and through reaffirmation of belonging to the people whom God had made. For Jesus to have used the word "body" in this context was to retain the corporate, social meaning of consuming the bread, and to associate his body with the power of God to create a new people, a new corporate body. For the Church to eat this bread now would be to eat a new people, to take in and thereby affirm association with God and the people through what has transpired in the flesh of Jesus. If for centuries the drinking of this cup had meant reaffirmation of the Sinai Covenant and Law, it became clear that Jesus had associated himself with God in the creation of a New Covenant, a new relationship to God through his total self-offering to God on the Cross. For the Church to drink this cup now would imply association with God in the response of Jesus to God by the power of the Spirit.

As noted in First Corinthians 10–11, for a brief time the ritual meal of the Lord's Supper preceded and followed a meal of satisfaction. However, as the mission of the disciples moved beyond Israel into Hellenistic culture, such a ritual meal had no ancient reference as a reaffirmation of a covenant that God had made. The Corinthians either rejected or found it difficult to grasp the radically corporate meaning of eating and drinking at the Lord's Supper, and hence that they ate and drank unto a deeper relationship with Christ and his members. Paul accused them of being guilty of insulting the Church because they imagined they could "get Christ" without recognition of the members of his Body around the table. This confusion led to the removal of the meal of satisfaction between the blessings, but *the basic structure of grace after meals would become the pattern of the Great Eucharistic Prayer to God in the hearing of the people.*[36] The elements and basic pattern of the Eucharistic Prayer include recitation of the great works of God, especially the life of Jesus who transformed the meaning of eating and drinking in that context, and prayer for the needs of the Church for the sake of the world. In this context, then, in eating the Body of Christ, the

36 Mitchell, *Cult and Controversy*, 24–27. Here it is noted that, precisely to preserve the meaning and integrity of the ritual meal, the meal of bodily satisfaction was necessarily dropped from the celebration of the Eucharist.

Church affirms its relation to God in and through the social Body, and in drinking the cup, gives itself to the New Relation to God revealed on the Cross.

CONCLUSION

William Shakespeare's *King Lear* (Act I, Scene I) demanded the subservience of his daughters as precondition for inheritance. Regan and Goneril adulate him, but Cordelia speaks the truth with affection, rather than with cowardice, duplicity or shallow loyalty. Lear banishes Cordelia, his "heart," to the admonition of Kent and the Fool to "See better."

Reformist feminists would have the entire Church "see better" that the Wisdom of God has assumed humanity in the person of Jesus of Nazareth.[37] The Wedding of Sophia with the matter of the universe through human flesh has sustained our grandmothers and mothers, all the women who have given us to the Church in this world, and countless thousands of women in religious communities, no matter what indignity has come their way. For, sometime near the end of his life, Jesus broke the silence after blessing God over bread, thereby restoring what God had intended in creation: a corporate body of women and men. He broke the silence after blessing God over the cup, thereby empowering every woman and man in this corporate Body to give themselves yet again to the New Covenant of God for the good of all the world. *The Spirit of God, by which Jesus broke the silence, is the Fire that heats the crucible in which the Church stands at Eucharist. Every issue surrounding the celebration of the Eucharist is judged by those words that broke the silence.* If the Church uncritically absorbed the patriarchy of the classical world, thereby muffling those silence-shattering words of Jesus, it is incumbent upon women to speak that truth with affection. The crucible is our hope.

Reformist feminists mean no disregard for the teaching authority of the Church, but choose to stand with the women of the Gospel standing at the Crib, at the Cross, at the Empty Tomb, and to pray for the healing

37 See Elizabeth Johnson, C.S.J., "Jesus-Sophia," chap. in *She Who Is: The Mystery of God in Feminist Theological Discourse* (New York: The Crossroad Publishing Company, 1992), 150–187.

of the heart of the Church. By the power of the Spirit who has ever surprised the Church

> We shall not cease from exploration
> and the end of all our exploring
> will be to arrive where we started [The Cross]
> and know the place for the first time.
> (The Fourth Quartet, "Little Gidding" T.S. Eliot)

Note: Portions of this reflection were presented to the national meeting of the Federation of Diocesan Liturgical Commissions at Providence, RI, October 1995.

Women's Struggle *for* Voice *as* Interruption *of the* Spirit

Mary Ann Hinsdale, IHM

Despite an overall dynamic understanding of revelation and tradition, the passages which deal with dogmatic development in Vatican II's *Dei Verbum* do not entertain discontinuity or interruption as a means of development. When speaking about development, *Dei Verbum* states: "This tradition which comes from the apostles develops in the church with the help of the Holy Spirit" (#8). But this activity of the Holy Spirit in the church, especially with respect to the development of doctrine, is construed mostly in an organic way that stresses continuity, coherence, and progression:

> For there is a growth in the understanding of the realities and the words which have been handed down. This happens through the contemplation and study made by believers, who treasure these things in their hearts (cf. Luke 2:19, 51), through the intimate understanding of spiritual things they experience, and through the preaching of those who have received through episcopal succession the sure gift of truth. For, as the centuries succeed one another, the church constantly moves forward toward the fullness of divine truth until the words of God reach their complete fulfillment in her. (DV, #8)

There are two footnotes which accompany these sentences from *Dei Verbum*. The first, from the Council, follows the statement concerning the development of tradition as the work of the Holy Spirit. It cites Vatican I's *Dogmatic Constitution of the Catholic Faith*, Chapter 4, "On Faith and Reason": Denz 1800 (3020). The second footnote, from the English translator, Walter Abbott, occurs a few sentences later and reads: "A description of the 'development of dogma.'" Here, Abbott alerts the reader to notice the order in which the "players" in the development process are named. He points out that the first medium of development are the faithful. First to be mentioned in this passage are the "believers," who, through "contemplation and study" and "the intimate understanding of spiritual things they experience," are compared with Mary, who treasured these things in her heart. "Preaching," which is done by "those who have received through episcopal succession the sure gift of truth," is mentioned second. Clearly, Abbott's wants to draw attention to the fact that doctrinal development begins first in the experience of the people.

Whether one agrees with Abbott's interpretation or not (though I do think his observation is significant), the basic understanding of development in this passage stresses constant, forward movement: "For, as the centuries succeed one another, the church constantly moves forward toward the fullness of divine truth until the words of God reach their complete fulfillment in her." (DV #8) Yet, besides the chapter on "Faith and Reason" from the First Vatican Council's "Dogmatic Constitution on the Catholic Faith," other footnotes to Article 8 go back even earlier, to the canons from the Second Council of Nicea and the Fourth Council of Constance, in order to stress that Vatican II's teaching on doctrinal development is, in fact, in continuity with what the church has always taught on this matter.

"BUNNYHOP ECCLESIOLOGY"

Here, in my opinion, we have an example of what a Baptist student in my seminary ecclesiology class some years ago called "bunnyhop ecclesiology":

> You Catholics, have a bunnyhop ecclesiology. The way you advance is like the bunnyhop. You are able to take two steps forward only after a lot of shaking to the right and left. Then, when you do you venture a small step forward, that small one is quickly retreated from with a hop backward. A quick disclaimer is made denying that the forward movement is anything new (i.e., "as Holy Mother Church as always taught"). Ironically, your insistence on the new teaching's continuity with the past covers over the *real* change; that which produces the advance, the step backward, actually disguises the momentum that is needed to propel you forward.

There is a complex argument hidden in this homey example that I think is quite profound. For example, compare what Vatican II says in *Dei Verbum* with the reference cited from Vatican I: "If anyone says that as science progresses it is sometimes possible for dogmas that have been proposed by the church to receive a different meaning from the one which the Church understood and understands: let him be anathema"

[Const. on Faith, Canon 3]. What lies at the heart of my student's frustration with so-called "bunnyhop ecclesiology" is the reluctance of official church teaching to admit that there could be error in past formulations of church doctrine. At the same time, there is also a failure to admit that the church has corrected past error by means of a new articulation. Thus, the question is raised: why cannot such an admission be made and recognized as such? Why cannot the inspiration of the Holy Spirit show itself in discontinuity as well as continuity?

Discontinuity in Church Teaching

Despite Vatican II's recovery of the eschatological motif of the "pilgrim church," its abandonment of triumphalism, and the deliberate intention on the part of John the XXIII to deal with error by means of the "medicine of mercy,"[1] a consistent reluctance perdures which refuses to recognize "moments of discontinuity" in past church teaching.[2] Again, in a footnote, Abbott notes that some Council fathers would have

1 The "medicine of mercy" refers to a demonstration of the validity of church teaching instead of the issuance of condemnations. See "Pope John's Opening Speech to the Council (October 11, 1962)," in Walter M. Abbott, general editor, *The Documents of Vatican II* (New York: America Press, 1966), 710–19.

2 This can be seen, for example, in *Gaudiam et spes*, which speaks of the role of the infidelity of the members of the church, both clerical and lay, over the course of many centuries:

> Although by the power of the Holy Spirit the Church has remained the faithful spouse of her Lord and has never ceased to be the sign of salvation on earth, still she is very well aware that among her members, both clerical and lay, some have been unfaithful to the Spirit of God during the course of many centuries. In the present age, too, it does not escape the Church how great a distance lies between the message she offers and the human failings of those to whom the gospel is entrusted.
>
> Whatever be the judgment of history on these effects, we ought to be conscious of them, and struggle against them energetically, lest they inflict harm on the spread of the gospel. The Church also realizes that in working out her relationship with the world she always has great need of the ripening with comes with the experience of the centuries. Led by the Holy Spirit, Mother Church unceasingly exhorts her sons (and daughters) 'to purify and renew themselves so that the sign of Christ can shine more brightly on the face of the Church.' (GS #43).

preferred to have here an enumeration of some of the concrete historical instances of infidelity to the Holy Spirit. Even the much-heralded invitation to scrutinize "the signs of the times" and to interpret them in light of the Gospel that was called for in *Gaudiam et spes* (#4) often presupposes an understanding of truth that is more static than dynamic.[3]

Ten years ago, historical theologian Walter H. Principe addressed the question of discontinuity in church teaching in an article in *The Ecumenist*. Principe's contention was that throughout history, both "theological reflection and the practical experience of the people of God who make up the church have led to modification of doctrines taught authoritatively by the Pope or curial officials."[4] In support of his thesis, Principe enumerated more than a dozen examples from Catholic church teaching which he labeled "moments of discontinuity."[5] Recently, a panel of theologians from the Catholic Theological Society of America questioned whether certain experiences in the present-day church, issues which the official *magisterium* of the Roman Catholic church has judged to be at odds with "the constant practice and tradition of the church," could be regarded as "interruptions" of the Holy Spirit. Arguing that the movement of the Holy Spirit cannot be confined to "organic development," the panel suggested that the Holy Spirit has interrupted the life of the church in the past. For example, both with respect to its teaching on the Jews and the option for the poor, the church's present teaching has undergone significant alteration, if not reversal.[6] If the action of the Holy Spirit can be recognized in these instances of discontinuity in church

3 For example, the following passage from *Gaudiam et spes*: "The Church also maintains that beneath all changes there are many realities which do not change and which have their ultimate foundation in Christ, who is the same yesterday and today, yes, and forever." (#10)

4 Walter Principe, "When 'Authentic' Teachings Change," *The Ecumenist* (July–August, 1987):73.

5 For example, the repudiation of slavery which took place only at Vatican II; the torturing of prisoners which was outlawed in the Papal states only in 1816; the change in the church's stance on religious liberty at Vatican II (and the rehabilitation of John Courtney Murray, who had been silenced and censured only a decade before); the modification of *nulla salus extra ecclesiam*; the broadening of Pius XII's understanding of the mystical body of Christ beyond identification with the Roman Catholic Church.

6 Gregory Baum argues that the "option for the poor" is in continuity with Catholic social teaching when it comes to its teaching on wealth, but in dis-

teaching, might not other situations of ferment in the church be similarly evaluated?[7]

There are, to be sure, the helpful ecclesiological principles invoked by Vatican II, such as the *sensus fidelium* (or as Vatican II preferred, *sensus fidei*; see LG, #12) and "reception" which could be used to justify characterizing "moments of discontinuity" as the interrupting work of the Holy Spirit. But the problem with these discernment processes (which is how I regard *sensus fidelium* or reception) is that they both comprise *retrospective* warrants for determining the movement of the Spirit. That is, they function by determining whether something *in the past* constituted authentic teaching or not. In the cases presented by the CTSA panel concerning the church's teaching regarding Judaism or the option for the poor, we are clearly dealing with matters that have been reversed in official church documents. One can acknowledge "discontinuity" simply by reviewing the magisterial statements of the past and comparing them with the positions taken at Vatican II or at Puebla. With respect to women's leadership in the church, or same-sex love, however, we face a different task. Here one must inquire whether it is possible to recognize the interruption of the Spirit in *present experience*, and if so, what would such recognition do to our understanding of the development of doctrine and the role of the Holy Spirit in that development?

WOMEN'S STRUGGLE

The specific example I want to consider in this article is women's struggle for voice and bodily agency in the church. Women's struggle is readily apparent from the feminist theologies and practices which have

---

continuity with it when it comes to power. Vatican II's repudiation of the Jews' guilt for the death of Christ reflects even a clearer example of change in official church teaching.

7 Gregory Baum, Rosann Catalano, Mary Ann Hinsdale and James Nickoloff addressed the topic "The Holy Spirit as Interrupting Agent in the Church" using the examples of the church's teaching concerning the option for the poor, the Jews, the role of women, and homoerotic love. See, *Proceedings of the Catholic Theological Society of America*, Vol. 51 (1996):247–48.

developed over the last thirty years. Based upon a communicative concept of justice,[8] Christian feminism advocates that church and society recognize women's full humanity as persons. The effort to enable women to realize their full humanity has given birth to a number of movements. Many of them, such as the Women's Ordination Conference (WOC) and Call to Action (CTA) are located in the USA, but have counterparts in many other parts of the world. Some of them actively work for change in the ecclesiastical practices that restrict women's ministry and leadership in the Roman Catholic church. Other movements, such as Women-Church Convergence or the Re-Imagining Community, are ecumenical and interreligious in membership. The goal of these organizations is to celebrate the sacredness of women's lives as reflecting the image of God in which all human beings are created. They do this by affirming multiple and multicultural dimensions of female experience, experimenting with the reconstruction of traditional Jewish or Christian symbols (i.e., the Passover Haggadah, the Lord's Supper, the cross). Rather than asking for inclusion into the present ecclesiastical or liturgical order these women are attempting to exercise their own agency, drawing upon the sensuousness of female bodiliness in order to affirm the biblical and ecclesial teaching of the equal humanity of women and men (Gal 3:28). More often than not, their imaginative efforts have been greeted with strong resistance and declarations which maintain that the tradition is "constant" regarding the "proper role" of women, whether that be in the home, in society, or in the church.

Despite this backlash, a many-colored quilt of feminist/womanist/mujerista theologies has arisen. These theologies are usually concerned with 1) critiquing the subordination of women and nature in the androcentric world-construction of patriarchy, a worldview which has treated not only women, but entire peoples and races as "other," and which has operated as the dominant paradigm in structuring the church; 2) retrieving women's hidden experiences of survival and salvation from the tradition; and 3) proposing imaginative visions of whole-

8 For the use of this concept I am indebted to Rebecca Chopp's discussion of communicative justice in *Saving Work: Feminist Practices of Theological Education* (Louisville, KY: Westminster John Knox Press, 1995). Chopp herself draws upon Iris Marion Young's *Justice and the Politics of Difference* (Princeton, N.J. Princeton University Press, 1990).

ness that do not integrate women into patriarchal structures, but enable both women and men to live in mutuality, sustained by the hope of the gospel.

FEMINIST EKKLESIA

What I propose for consideration is that feminist/womanist/mujerista theologies, as well as the practices of what Elisabeth Schüssler Fiorenza calls "women-church"[9] (and Rebecca Chopp calls "feminist *ekklesia*"[10]), represent an "interruption of the Holy Spirit." Furthermore, I want to suggest that both the theologies and practices of religious feminism present the church with a dialogical process, the goal of which is to discern the interruptions of the Spirit in the present, and which should not simply have to depend on "the test of time"—when it is too late for those who have been silenced or fired (this usually being the price paid for prophetic interruption). By examining the experience of the diverse women-church, or feminist *ekklesia* movements, which have in common the recognition of the full-humanity of women in the church and society, I hope to show that the experience of these groups provides us with a concrete model of ecclesial discernment in which "truth" and "error" do not need to be seen as irreconcilable. Here, I believe, the theologies and practices of religious feminists have much to offer the church by way of presenting models of sustained conversation and dialogue that are based upon an eschatological understanding of truth. Although this is simply an intuition at this point, I further believe that closer study of the models of communicative justice that are emerg-

9 Elisabeth Schüssler Fiorenza, "For Women in Men's Worlds: A Critical Feminist Theology of Liberation," in Claude Geffre, Gustavo Gutierrez, and Virgil Elizondo, *Different Theologies, Common Responsibility: Babel or Pentecost?* (Edinburgh: T & T Clark, 1984), 37. Schüssler Fiorenza speaks of the *ekklesia* of women, not as a separate strategy, or to mythologize women, but "to make women visible as active participants and leaders in the Church, to underline women's contributions and suffering throughout Church history, and to safeguard women's autonomy and freedom from spiritual-theological patriarchal controls." She proposes that just as there is a "church of the poor" that does not prevent the theological vision of the universal Catholic Christian church, so too is it justified to speak of women-church as manifestation of the universal Church.

10 Chopp, 45–71.

ing from the *ekklesia* of women could be a helpful resource for refining both the theory and the praxis of such traditional ecclesial discernment processes as *sensus fidelium* and reception.[11]

A communicative model of justice is not simply based upon the Enlightenment concept of "rights"—though it is that—nor is it based upon a paradigm of "distribution."[12] Communicative justice is that which defines the nature and mission of the feminist *ekklesia*. It involves having a voice, being able to participate in that which affects one's destiny.[13] As Rebecca Chopp puts it so well, "in the feminist practice of theology, justice is central to the braiding together of ethics and epistemology in the formation of new meanings and functions of symbols and the development of new discursive practices."[14] What I am calling "women-church" can perhaps best be illustrated by looking at three concrete embodiments that capture the main features of this "*ekklesia* of women," events that appear to be an "interruption" (some would say, "disruption") in the life of the Christian community: the Women's Ordination Conference, Women-Church Convergence, and the Re-Imagining Community.

11 Ibid., 62–71. Another resource whose practice of communicative justice may be helpful in this regard is that of Roman Catholic congregations of women religious who, in the face of increased theological pluralism among their own membership, have committed themselves to processes of "sustained interaction," rather than severing connections with each other by splitting into separate communities.

12 Basing herself upon the work of Robin Lovin, Rebecca Chopp explains that the distributive paradigm of justice "concerns the distribution of goods, rights and responsibilities among the members of a community." It is essentially "a nonconstitutive act of community; that is, it doesn't make up the real web of community, but distributes the secondary affects and concerns. In modern American society, justice is primarily managed as a matter of the welfare state. For mainline denominations, justice is a matter of charity, or distribution of surplus goods to the poor." See, Chopp, 62.

13 Although the term, "communicative justice" is not used (and this term should not be confused with "commutative" justice), this essentially is stressed in the concept put forward by the U.S. bishops in their 1986 pastoral letter on the economy, "Economic Justice for All." In their endorsement of the principle of participation, the bishops consider the main deprivation of poverty to be "an inability to influence decisions that affect one's life" (Cf. #188). This kind of "voice" is what women are denied in the church today.

14 Chopp, 106.

The Women's Ordination Conference is an organization which has been in existence for over twenty years. Despite recent strategic differences among its members, WOC is an example of what Rebecca Chopp calls a "place of grace." Feminist ecclesiology does not assume that "where the institution is, there is the church," but "seeks to name the church as that place which opposes patriarchy and envisions new forms of flourishing.... in terms of the denunciation of sin and the annunciation of grace."[15] Chopp's description of "*ekklesia*" as a "place of grace" explicitly eschews identification with separatist groups or with a church for women only. "Rather, the feminist *ekklesia* is that place of God's redemptive presence where women and men can be emancipated from sin and transformed into freedom. In this sense, the *ekklesia* is defined by the presence of the Spirit."[16]

Women-Church came into existence in the mid-1980s. It is, in the words of Jane Redmont, "neither a new church nor, by the traditional definition, a schismatic group."[17] Women-Church sees itself as a place on the margins, where women are able to practice religious self-determination. Expanding the notion of sacrament, Women-Church gatherings affirm the celebrations created by and out of women's experience as sacramental. Like other "Christian base communities," both WOC and Women-Church have tried to create "discursive spaces" that enable women to denounce sin and to announce grace. Both movements function to work against sin inside the church and in society.

A third example of the *ekklesia* of women that I want to discuss is what has come to be known as the "Re-imagining Community." The Re-Imagining Conference, an ecumenical event held in Minneapolis in November 1994, became the victim of distorted (often vicious) reporting, ecclesial backlash, and outright censure.[18] The purpose of the confer-

15 Ibid., 52.

16 Ibid., 53.

17 Jane Redmont, *Generous Lives: American Catholic Women Today*. (New York: William Morrow and Co., Inc., 1992), 283. See also, Miriam Therese Winter, "The Women-Church Movement," *The Christian Century* 106 (1989): 258–60; Mary Hunt, "Spiral, Not Schism: Women-Church as Church," *Religion and Intellectual Life* 7 (1990): 82–92; Rosemary Ruether, *Women-Church: Theology and Practice of Feminist Liturgical Communities* (San Francisco: Harper and Row, 1985).

18 The reaction of official ecclesiastical bodies has been varied. Neither Women-Church nor WOC are officially connected with church bodies, but

ence was to theologize and ritualize Christian community using women's voice and bodily agency as the locus of theological reflection. The meeting was demonized by conservative publications and groups within both the Methodist and Presbyterian churches.[19]

Participants in the gathering were accused of "goddess worship" for invoking the female personification of Wisdom during the prayer rituals. Yet, as one participant noted, the conference liturgies were distorted by "ripping image after image out of context, yoking together the most sensuous metaphors in order to create an aura of pornography." For example, a line which especially produced outrage was the following: "We are women in your image. With nectar between our thighs, we invited a lover, we birth a child." Although this mixing of sensuality and spirituality can be found in the Song of Songs ("My beloved is to me a bag of fragrant herbs that lies between my breasts; I am faint with love"; "My beloved is knocking... I arouse to open")—and perhaps its biblical poetry is to be preferred—one would be "hard pressed to judge which is more graphic," commented theologian Catherine Keller.[20]

Womanist theologian Delores Williams, a professor at Union Theological Seminary, was severely castigated for her remark concerning the effect of the doctrine of atonement on women. Her oft-quoted remark, "I don't think we need a doctrine of atonement at all. I don't

---

the membership of both groups is largely Roman Catholic. This is particularly true of WOC, whose membership includes many Catholic women religious, theologians, and parish ministers, and which has functioned as an independent "lobby group" for women's ordination in the Catholic church. However, some might see the recent letter of John Paul II, *Ordinatio Sacerdotalis*, which asserts that the prohibition of women priests is to be held as "definitive" and precludes further discussion, as a censure to any person or group advocating women's ordination. The "Re-Imagining" conference, though not officially sponsored by any one denomination, received considerable financial backing from the Methodist and Presbyterian churches (see, "A Controverted Conference," *The Christian Century* 111 [February 16, 1994], p. 160). Mary Ann Lundy, who served as director of the Presbyterian Women's Ministry Unit and liaison to the conference planning committee, was forced to resign her staff position on the General Assembly Council of the PCUSA.

19 Catherine Keller, "Inventing the Goddess," *The Christian Century* 111 (April 6, 1994): 340–42.

20 Ibid., 341.

think we need folks hanging on crosses and blood dripping and weird stuff," was repeated in *Good News* (a conservative Methodist publication) and the *Presbyterian Layman*, without any effort to contextualize it. Her words may have been ill-chosen, but the critiques make no mention of the feminist/womanist scholarship that warns how an exclusive emphasis on the salvific nature of Jesus' self-sacrifice creates "unhealthy roles of self-abnegation" and serves to "shackle women in abusive relationships." Williams' challenge was simply condemned outright as destroying Christian faith.[21]

Finally, accusations were made that the conference affirmed homosexuality by applauding the presence of lesbians. *The Christian Century* reported that an unscheduled gathering of about 100 lesbians on the dais received a standing ovation from the audience.[22] Interestingly, a similar occurrence took place at the Women-Church "Weavers of Change" conference in Albuquerque. According to *Presbyterian Layman*, the mouthpiece of the Presbyterian Lay Committee, an evangelical caucus in the PCUSA, the affirmation of lesbian lovemaking was a recurrent conference theme.

21 See David Heim, "Sophia's Choice: Wisdom and Conversation," *The Christian Century* 111 (April 6, 1994), 339. Williams' acclaimed book, *Sisters in the Wilderness: The Challenge of Womanist God-Talk* (Maryknoll: Orbis, 1993), reflects upon the experience of black women in surrogacy roles in order to re-assess Christian understandings of redemption that stress crucifixion as salvific. Apparently neither this work nor others which treat similar themes were known to her detractors. For other works on this topic see, for example, Joy Bussert, *Battered Women: From a Theology of Suffering to an Ethic of Empowerment* (New York: Division for Mission in North America, Lutheran Church in America, 1986); Joanne Carlson Brown and Rebecca Parker, "For God So Loved the World?" in *Christianity, Patriarchy and Abuse*, eds Joanne Carlson Brown and Carole Bohn (New York: Pilgrim Press, 1989); Christine Gudorf, *Victimization: Examining Christian Complicity* (Philadelphia: Trinity Press International, 1992). Since the Re-Imagining Conference, several responsible articles appraising the feminist critique of atonement theology have appeared. See, for example, Leann Van Dyk, "Do Theories of Atonement Foster Abuse?" *Dialog* 35 (1996): 21–25 and Thelma Megill-Cobbler, "A Feminist Rethinking of Punishment Imagery in Atonement," *Dialog* 35 (1996): 14–20.

22 "A Controverted Conference," *The Christian Century* 111 (February 16, 1994): 160.

It was to be expected that conference participants, church officials, and a variety of women's organizations responded to these attacks. Unfortunately, according to David Heim, many of the public defenses simply returned polemic for polemic.[23] Yet, other respondents attempted to focus on the issue of women's voice. Elizabeth Dodson Gray defended the conference, saying that the real issue was "the reality of women finally standing up in the Christian tradition as equal participants in "the naming game."[24] Patricia Humer of Church Women United, said: "Men need to silence this kind of thing in order to be in control." Catherine Keller echoed this sentiment, pointing out that "the power of women can appear overwhelming to men—and some women—who are on the defensive. As women gain power, men who feel that their own power is diminished are tempted to stress the 'her' of heresy... the strength of the backlash is testimony to the power of our accomplishments..."[25] Eight hundred United Methodist Women, including deaconesses, laywomen, and college faculty, issued a statement supporting the conference. Called "A Time of Hope—A Time of Threat," the signers accused the naysayers of: 1) refusing to acknowledge the positive relationship between sexuality and spirituality; 2) accusing feminist, womanist, and other women theologians of departing from the Christian faith; 3) homophobia, revealed by their verbal violence against lesbians; and 4) creating a climate of witch-hunting, name-calling, and fear that destroys Christian community.[26] United Methodist Bishop Susan Morrison, who attended the news conference releasing the statement, drew laughs when she expressed her frustration with the attacks from the conservative Christian groups: "We can talk about God as a mineral—'Rock of Ages'—but when we speak of God as feminine, it creates a crisis."[27]

In his analysis of the Re-Imagining controversy, David Heim points out that a necessary stage of dialogue is "to be able to state the other

[23] Heim, 339.

[24] Elizabeth Dodson Gray, "Interpreting the Furor Over the Re-Imagining Conference," Press release of February 25, 1994.

[25] Keller, p. 342.

[26] "Women's conference and a 'theological crisis,'" *The Christian Century* 111 (March 23–30): 306–07.

[27] Ibid., 307.

person's position in terms that the other will find reasonably accurate."[28] He notes that careful attention to the opponent's point of view is difficult, tedious, and often "less than stirring." He even ventures that this is why careful ecumenical discussions on faith and order attract a limited audience. He endorses "imagining," but maintains that one must be ready to test it against Scripture, tradition, and Christian experience. For Heim, attainment of true wisdom involves conversation:

> This approach will not appeal to those who would banish imagination from theological thought, or who think tradition is fixed and settled, or who reject out of hand all revisions of ancient doctrine. Nor will it attract those who regard the tradition as so corrupt that it must be entirely re-imagined, or so bankrupt that continuity with it is not prized.[29]

Nevertheless, he continues, "honest and charitable debate and criticism are necessary if Christians are to understand, judge and act on matters that demand the church's attention."[30]

Women's Struggle for Voice as Inculturation

The "Re-Imagining" conference is simply the most recent example of how the church is being challenged by women who are insisting that their experience as women be taken seriously as a locus for Christian theology. In many ways, this event raises yet again the perennial question of how Christianity can remain "catholic" in the face of diverse cultural expressions.[31] Seen in this light, women's struggle for voice and

28 Heim, 339.

29 Ibid., 340.

30 Ibid. In fact, such conversation has already begun to take place. The Office of Theology of the Presbyterian Church (USA) issued a statement, "Evaluating 'Re-Imagining,'" which was published in *The Christian Century* 111 (April 6, 1964). Elizabeth Bettenhausen's article, "Re-Imagining: A New Stage in U.S. Feminist Theology," in Ofelia Ortega (ed), *Women's Visions: Theological Reflection, Celebration, Action* (Geneva: WCC Publications, 1995): 90–101 and the book edited by Nancy J. Berneking and Pamela Carter Joern, *Re-Membering and Re-Imaging* (Cleveland, OH: The Pilgrim Press, 1995), containing both affirmative and negative evaluations of the conference, are definite moves toward such conversation.

31 I am using "catholic" in the sense of "universal" here.

bodily agency in the church can be approached in terms of the inevitable tension that exists between catholicity and inculturation.[32]

Traditionally, it has been the *sensus fidelium*, the participation of the baptized "faith-full" in both the formulation of and witness to what constitutes "the faith," that has been the determinant of catholicity.[33] According to this doctrine, because the Holy Spirit is its foundational origin, the "sense of the faithful" is infallible, and preserved by the Spirit from error in matters necessary to revelation.[34] However, as Orlando Espin has pointed out, what is problematic about this doctrine is that it is a "sense," an intuition. As such, it is never discovered in some kind of pure state:

> The *sensus fidelium* is always expressed through the symbols, language, and culture of the faithful and, therefore, is in need of intense interpretive processes and methods similar to those called for by the written texts of tradition and scripture. Without this careful examination and interpretation of its means of expression, the true "faith-full" intuition of the Christian people could be inadequately understood or even falsified.[35]

By suggesting that women's struggle for voice is an invitation for the church to embrace the challenge of inculturation and the opportunity to

32 For an excellent discussion of this tension, see Walter H. Principe, "Catholicity, Inculturation, and Liberation Theology: Do They Mix?" *Franciscan Studies* 47 (1978): 24–43 and Paul Crowley, "Catholicity, Inculturation and Newman's *Sensus Fidelium*," *Heythrop Journal* 33 (1992): 161–74.

33 See *Lumen Gentium*, #12. In addition to Crowley's article mentioned above, see the following for recent discussions of *sensus fidelium*: John Burkhard, "*Sensus fidei*: Meaning, Role and Future of a Teaching of Vatican II," *Louvain Studies* 17 (1992): 18–34; Edmund J. Dobbin, "*Sensus Fidelium* Reconsidered," *New Theology Review* 2 (1989): 48–63; Patrick J. Hartin, "*Sensus Fidelium*: A Roman Catholic Reflection On Its Significance for Ecumenical Thought," *Journal of Ecumenical Studies* 28 (1991): 74–87; Michael McGinniss, "*Sensus Fidelium*, USA: Laity and Church Structures For the Future," *Listening* 25 (1990): 71–85; Orlando Espin, "Tradition and Popular Religion: An Understanding of the *Sensus Fidelium*," in Allan Figuero Deck, ed., *Frontiers of Hispanic Theology in the U.S.* (Maryknoll, NY: Orbis Books, 1992): 62–87.

34 Orlando Espin, "Tradition and Popular Religion," 65.

35 Ibid.

foster a true catholicity, I do not wish to deny the necessity of discernment that David Heim called for in the "Re-Imagining" controversy. As Espin points out, interpretation and discernment is always needed in determining *sensus fidelium*. Thus, the church must try to ascertain both the authenticity of the intuitions as well as the appropriateness of the expressions.[36] Such a process will call for "confrontations": with Scripture, with other written texts of tradition, and with the historical and sociological contexts within which the "intuitions of the faithful" and their means of expression appear.

It is this third confrontation that I believe is most neglected and the one which is most likely to present what might be judged an "interruption," namely, whether any theological reflection on the experience of faith (taking note of particular cultural, political, economic, racial and gender contexts) can ever exhaustively define what is true and fundamental for the whole human race. Here, in the context of assessing Hispanic popular religion as reflective of the *sensus fidelium*, Espin proposes that if such intuitions of the Christian community can only be expressed through culture-bound means, then "it is possible that the same intuition could be communicated by different Christian communities through different cultural means."[37] In my estimation, this means that acceptance of inculturation not only leads to theological pluralism, but also that it ultimately will be perceived as the Holy Spirit "interrupting" a certain cultural hegemony. Although the examples of women's struggle for voice as full members of the Christian community recounted here are primarily anecdotal, I believe they represent an example of the Holy Spirit's interruption in the church. I do not single out women's struggle for voice because I think it is the only example of the Spirit acting as an interrupting agent in the church today, but because such a struggle represents an attempt of ordinary people—mostly ordinary women—to respond to what they consider the prompting of the Spirit. Above all, it is the Holy Spirit who affirms women's experience as persons created in the image and likeness of the divine. Especially for Catholic women, the consistent claim of the last three magisterial documents on women's

36 Ibid.

37 Ibid., 66.

ordination[38] contradicts both women's own experience of the Spirit as well as the traditional doctrine of the *imago Dei*.[39] The most recent statement, asserting that the prohibition of women's ordination is "founded on the written Word of God, and from the beginning constantly preserved and applied in the Tradition of the Church," and requires "definitive assent" because it has been set forth infallibly by the ordinary and universal Magisterium, only serves to heighten the question.

It seems to me that a more comprehensive understanding of the role of the Holy Spirit in the development of doctrine, a role which includes "discontinuity" and "interruption," as well as "organic development" and "constant teaching," would lessen this tension between inculturation and catholicity. In this regard, the *ekklesia* of women serves as a prophetic impulse, one which provokes the church to look at the inadequacy of its "bunnyhop ecclesiology" and lack of a concrete discernment process for determining the *sensus fidelium*.

I am often asked, and indeed, sometimes ask myself, why so many women still choose to stay in the church, given the suffering and abuse to which they are subjected? I believe that one reason that women choose to "stay in the conversation," is because they have *already* experienced a Spirit who "interrupts" in order to lead us to the truth. Like so many marginalized groups, women have experienced church in what Elisabeth Schüssler Fiorenza calls the *counter-public sphere*: "an oppositional space to the dominant public sphere in which the critique of patriarchy is generated and feminist visions and interests are articulated."[40]

38 The claim is that the inadmissibility of women to priestly or episcopal ordination is a matter of "the Church's constant tradition" (*Inter Insignores*, 1977) or "the constant practice of the Church which has imitated Christ" (*Ordinatio Sacerdotalis*, 1994). The most recent statement is the Congregation for the Doctrine of the Faith's *Responsum ad Dubium* (1995), which sought to clarify any doubts concerning the "definitive" nature of the previous statements.

39 It is beyond the scope of this essay to review all the arguments made in the various documents prohibiting women's ordination in the Catholic church. One of the more "notorious," however, is the insistence upon a theological anthropology of sex complementarity which accords certain roles to men and women on the basis of biological sex. On this basis, women may not act *in persona Christi* because they are not biological males.

The "counter-public" sphere of church is the space where women have always participated in decision-making and deliberation about their lives. By providing a space to recall these experiences and to imaginatively create new processes for "truth telling," the *ekklesia* of women presents the church with a model of community based upon communicative justice. This is an "interruption of the Spirit" that the church should welcome as a valid intuition of the faith-full.

40 Chopp, 63.

Male Identity, *the* Women's Movement, *and the* Need *for* Dialogue

Ronald Barnes, SJ

Part I: Gender Relations in a Time of Change: A Male Perspective.

The contemporary period continues to be an eventful time in the ongoing drama of the sexes. Women and men around the globe are facing major challenges to the way they have learned to think, feel and behave toward one another for as long as we can remember. The social, political and economic structures that have long governed our lives and interactions for centuries are in a state of flux. Changes are taking place that will alter the roles and relationships that have prevailed between the genders for generations and will undoubtedly have a profound effect on the way we will inhabit the earth together far into the future. The intent of the present paper is to explore some aspects of the experience of being male during this period of rapid and extensive change in gender relations.

On balance, the contemporary period is not a particularly felicitous time to seek the Holy Grail of authentic masculinity. While it may be true that women continue to trail men in many measures of gender equality, such as economic power, political influence, or physical safety, this situation is rapidly changing, especially in the developed countries. In fact, recent studies suggest that in Europe and North America the comparative status is rapidly reversing. Women are overtaking and surpassing men in almost all areas of academic success, while fields that traditionally attract women are expanding and those of men are declining. Already unemployment projections for the future are much more dire for men than they are for women.

Psychologically, as well, men are in a markedly defensive position. In the demand for change, women have a manifestly legitimate cause and clear principles of redress. Around the world there is an ascending pride in the goodness of being female and the conviction that feminine characteristics are often more in keeping with the needs of human society and the future health of the planet than typically masculine characteristics. Aside from ideological questions such as who is or is not a feminist, or what differences there may be among feminist positions, most women see the recent change as part of a positive development that has already achieved many tangible benefits, and promises to bring many more. Their response is therefore generally enthusiastic, cooperative and initiatory. They have a sorority of solidarity with a mutual purpose. In the language of trendy sports commentators, women are on a roll!

For many men however, the challenge tends to be much more ambiguous. While some men do perceive the situation as a promising opportunity to develop a more wholesome paradigm for human society in general, many more find themselves disconcerted and defensive. Caught off guard by the intensity and extent of the revisioning taking place around them, they find themselves at best cautiously acquiescing to essential changes, especially those legislated by governments or institutions, while numbly resisting others. They remain uncertain about the final goals and parameters of the process, and uncertain about their own place in a rapidly changing environment which fiercely challenges their privileged status in society and is highly critical of traditional male values and behaviour. For many the entire adjustment appears to be undermining their social status in society, and even threatening their sense of personal identity and security.

The posture of following behind the women's movement and responding to pressure points, largely without a coherent program of their own, has not only been costly to male self-esteem, but has also hampered effective dialogue between the sexes and served to foster one-sided analysis and exaggerated accusations of male shortcomings. It would seem important at this time, then, for men to take a more responsible and educated role in the search for new models of gender relations and social structures that would accommodate both sexes. Not only is such participation necessary for their own welfare and development, but a successful resolution of the adjustment between the genders would help to improve the quality of life on the planet for both men and women and reduce the tensions and injustices that currently plague our societies.

However, it is essential that men appropriate their own experience of masculinity so that any new roles and models for being male that emerge will genuinely express their own reality, rather than be fashioned solely in response to the demands of women, a solution which would ultimately satisfy no one. For this to take place however, men need to know something about how the present patterns of being male and female together have evolved over the centuries, why many of these patterns may no longer be acceptable to large numbers of the population, and how men themselves can begin to cope with the challenge of seeking new forms of self-understanding and relationships. Perhaps we can begin by acknowledging some of the peculiar difficulties involved in growing up male today, as well as the pain and confusion that frequently accompany the process.

Problems with Being Male Today

Gender is an extremely deep and intimate dimension of human existence. A person may have red hair or be adept at cards, but neither of these characteristics defines one's sense of self in any way comparable to that of gender. We are reminded of being female or male constantly during all hours of the day, whether we are watching TV, talking to someone, or merely walking among others on the sidewalk. Whenever we are awake we are aware of ourselves at some level as male or female, and any significant adjustment in how we are expected to understand and live this out is bound to effect us very deeply. And it can be said with some truth that at the moment, men are definitely being asked to accept a reassessment of their place in the scheme of things which closely effects their traditional self image, and the process is proving to be a painful and challenging experience.

Living with the Feminist Critique

It is not uncommon today for men to feel themselves subjected to a fairly steady flow of corrective pressure and implicit criticism of a rather generic kind. One of the disconcerting aspects of dealing with the women's movement for most men is that it quite legitimately appeals to their sense of fairness. In calling for a more equal and fair distribution of the goods and influence of society, women are merely calling for their due, a fact that is virtually impossibly for any honest and thoughtful person to deny. A simple glance at the daily newspapers demonstrates that men still control most business ventures and receive the preponderance of financial remuneration, while women suffer the most as battered persons or sex abuse victims, receive inequitable remuneration for similar work, endure the greatest burden in caring for children and house upkeep, and suffer the greatest poverty and loss of earnings in separation and divorce, to name only a few categories.

The 1992 study of UNICEF "On the State of the World's Children," refers to the "apartheid of gender" as one of the greatest reasons for child mortality and poverty worldwide.[1] It documents that the female population in the underdeveloped world is now doing the majority of its work, producing three quarters of its food, working twice as many hours a day as men, and being rewarded with one third of its food, financial remuneration, and educational opportunities, while enduring the greatest amount of dislocation, refugee status, disease, hunger and starvation.

In their analysis of the causes of this situation, feminists point to hierarchical patriarchy as the primary culprit. They argue that our social structures have long been organized along the lines of masculine leadership, paternal power and the subordination of women and children to this system. From this perspective, men are not only the beneficiaries of unfair advantage, but also, in effect, oppressors. Most feminist authors stress that it is the structures which must be changed, and that it is not men as such that are the object of their critique. Nevertheless, feminism has become a lightning rod that conscientizes men to a recognition of the inequity of the present system and raises uncomfortable questions about their own complicity in supporting and benefiting from it.

As an example of the increasing social pressure on men in general, the reaction to the sexual deviance and random violence of some men has led to a demand for greater safeguards directed toward all males. Many school districts now have strict rules against male teachers touching children at all, even requiring them to call a female teacher when the children fall and harm themselves in the school yard. Little things like holding a door for women in a public building is now an awkward and uncertain gesture, and men are advised to cross the street rather than walk behind or overtake a woman on the streets at night. As a result a sensitive or conscientized man is often abnormally aware of being male and frequently feeling apologetic about his presence.

As with any large and popular movement in which people are

[1] "The State of the World's Women 1985," Compiled for the United Nations by New Internationalist Publications, Oxford, U.K., 1985.

endeavoring to correct years of injustice and hurt, the women's movement is naturally accompanied by a certain amount of ideological rigidity and exaggerated demand which can be both wearisome and unfair. Depending on their working environment, men can be subject to a steady atmosphere of kindly correction or conflicting expectation that could make the most self-confident person unsure and occasionally resentful. Much of this may be inevitable and unavoidable for the present, and the discomfort may be small compared to the fears and injustices that women have been suffering for ages, but it does heighten the sense of uncertainty and sometimes of exasperation that often goes with being male today.

Even in those rare settings where men and women are able to explore gender issues together with calmness and respect, one generally finds that women are much better informed and firm in their convictions, while the men are tentative and exploratory. Often hesitant and publicly deferential towards the expressed views of women participants, men normally entertain a good deal of private reserve about many of the general assumptions made about masculine experience and characteristics, as well as about much feminist analysis of the past. At any rate, most discussions in which men are gradually led into an honest exchange will reveal a good deal of uncertainty about what would actually constitute a satisfactory and distinctive masculine profile or spirituality today.

The Loss of Vibrant and Acceptable Masculine Symbols

A major challenge to male confidence in the contemporary setting is the loss of esteem for traditional male symbols. Claiming a "feminization" of institutional religion, Patrick Arnold once made a plea for the resurgence of a more virile, heroic or "dangerous" masculine spirituality akin to what he sees as the soul-stirring and iconoclastic mission of Jesus.[2] Yet the typical symbols of masculinity such as hero, king, warrior, explorer, protector, have commonly become associated with violence, militarism, power, exploitation, fascination with technology to the detriment of the environment, and the lack of human compassion and affection which have marred so much of our century. It is thus

2 Arnold, Patrick M. *Wildmen, Warriors, and Kings: Masculine Spirituality and the Bible*. New York: Crossroads, 1991.

difficult for sensitized men to know how to appropriate these traditionally distinctive symbols, and often the only alternative seems to be highly feminine substitutes.

Carl Jung's well known claim that the psychic illness of many people in the modern world stems largely from a loss of contact with the life-giving archetypal symbols that enable us to negotiate the depth dimensions of existence, is very much to the point here. The rapid overthrow of familiar male images makes it difficult to formulate compelling symbols for young boys to emulate and from which adult men can derive guidance. The human spirit needs meaningful symbols to feed upon and western men are particularly bereft right now.

A PRECARIOUS SENSE OF IDENTITY AND MEANINGFULNESS

Perhaps reflecting the dislocation of masculine archetypes, a number of authors and professionals point to a disconcerting spirit of malaise and lack of purpose among contemporary male populations. Social workers and psychologists speak of an emptiness in the outlook of many young male clients who see little worth or meaning in their lives, leaving them prone to boredom or aimless violence. The American poet Robert Bly and others report a demand for all-male workshops where participants seek close personal contact with other men in an effort to discover what makes them distinctive as men, or to restore a sense of rootedness.

Recent studies have indicated that joblessness among poorly educated young men in Europe and North America is becoming endemic, and looks as though it may become a permanent feature of the new age. Spiritual director William O'Malley complains about "a soullessness" or a "God sized void" among the young men with whom he ministers, leading to an addiction to loud music, graffiti, gang violence, aversion to commitment, or blind adherence to grinding routine that substitutes for the loss of self that many of them feel.[3]

In short, then, there are signs of a marked reduction in masculine contentment and self-confidence, and a lack of clarity about what constitutes a distinctive male role. Many men are uncertain about what it means or should mean to be a man today.

3 O'Malley, William, "The Grail Quest: Male Spirituality," *America*, 166/16 (May 9, 1992): 405–6.

Part II: Industrialization and the Masculine Predicament

It would be helpful to recall that the arrival of the modern industrial age had a major influence in the shifting fortunes of masculine identity in our era. In particular, the transition from a more rural and pastoral society to an industrial and modern one has had an immense impact on the different ways that people live their lives today than they did in earlier centuries. It has been estimated that in the early middle ages up to ninety five percent of the population of Europe was involved with the land for their livelihood, compared with about seven percent today, a tremendous transformation!

As that massive transition from a rural and pastoral society to a modern one was taking place, people were drawn off the land and into cities, and public transportation arrived to enable people to live in one part of the city and work in another. Since much of this labor was in heavy industry, men were drawn into mines, steel mills, and ship building yards, so that the pattern was established in which men would be mostly employed outside the home and women would be in charge of the domestic setting. This effectively withdrew the male population from the home during the day and exaggerated the separation of the roles of men and women in society, both of which have resulted in a less than ideal setting for the rearing of children, perhaps young boys in particular.

The Absence of Available Male Role Models: An Illustration

For a basic paradigm let us turn to the coalminer in D.H. Lawrence's novel, *Sons and Lovers*, which portrays the fortunes of a small laboring family in 19th century England. The novel is set in a small English village where the father was part of the work force in a coal mine. The mine could as easily have been a shipyard, a steel mill, or a construction company, and with adjustments, the model could apply to the lives of many people employed by large corporations today. In the story, the father followed his own father into the mines with a minimum of education and little choice over the times or conditions of his labor. Exhausted by the daily grind and without access to revitalizing recreation, he becomes more and more sullen and less and less able to take part in the inner life of the family. His life is circumscribed by the sound of the horn that summons the men of the valley to the different work shifts, and the need to restore his body with sleep and his spirit with the few hours of banter and oblivion provided by the pub. His importance to wife and children becomes more and more defined solely by the money that remains from his drinking and the need to avoid provoking his sullen rage, which in most cases is effectively checked by fear of his wife's resolute inner strength. There is little to stimulate his personal development and the harsh conditions of the times insist that he take whatever work shifts are available if the family is to eat.

Meanwhile his children know nothing of the coal mine, or the equivalent office or factory where the father works, so that his daily activity remains a mystery to them. Having little in common with him or his world, they find him more and more anachronistic as they grow older. Fearful of his temper, and aware of the fitful resentments between him and their mother over his drinking and waste of money, they grow more and more indifferent and ashamed of him. His wife, a talented and determined person, wants nothing more for her sons than that they become educated and escape the grip of the mine by qualifying for a better life. The father senses this and initially struggles to hold the children, but he has no leverage and ultimately succumbs to drinking and morose, stubborn silence sprinkled with bouts of crudity or threats of violence around the house. The eldest son, being a favorite of the mother and strongly under her influence, develops a taste for the arts and advances to escape the drudgery of heavy labor by becoming a clerk. But he is so estranged from the masculinity exhibited by his father, while being held under the vaguely erotic spell of his mother, that he spends his adult life flirting with a long time relationship to a girl he never marries, while engaging in a series of passing relationships along the way.

For the duration of his adult life the father continues to leave home soon after he wakes in the morning and returns at night too tired and peevish to engage the wife or children in any personal manner. His life is dominated by the narrow and specialized world of his occupation, and he has neither the opportunity nor the energy left to develop broader aspects of himself through leisure, travel, education or casual conversation. Philosophical questions may rove in his imagination occasionally but they are truncated by the ever sounding whistle of the mine calling him to work, the drudgery of the work, fear of being laid off, constant exhaustion and the drivenness of whatever relief he does find in escaping to the pub.

In short, the father offers very little that is exciting and admirable to his sons as a model of what they would like to be as adults. He is not someone whom they can consciously and proudly emulate in their search for what it means to be a man. Even today, both laborers and professional men often find themselves similarly trapped in a narrow and unrewarding work environment which becomes all-embracing, leaving minimal time or energy for their growing family. In fact, available leisure time for both men and women has dropped significantly even within the last fifteen years, and it is not uncommon for couples in the United States now to hold four jobs between them, while their adolescent children work part-time as well.

And so, while the harsh physical realities of Lawrence's story are not quite as prominent in the developed countries as they were in the past, the pattern which many men still present to their children during their growing years remains problematic. Robert Bly refers to it as the "absent father syndrome" which is passed on from generation to generation, leaving young males without the comfort and benefit of a sturdy and familiar model of what it means to be a man during growth years. This is especially difficult since anthropologists tell us that male identity formation is much more tenuous and variable than the female. The passage into male adulthood has always been a more socially constructed and "costly goal" initiated by specific and dangerous tests such as the killing of a dangerous animal or navigating a river or open ocean crossing, by which costly feat the young man is then formally recognized as a man. Woman, by contrast, more easily and naturally evolve into recognizably mature adults, and are accepted as such without quite the same need for proof of adulthood.

The lack of convincing models and an absent or distant relationship with the father can be a special hardship for the young male. And the consequent distance from their wives and children in society hardly enhances the quality of life for adult men themselves, many of whom also suffered a thin relationship with their own fathers. Painful regret over weak emotional contact and lack of quality time with their fathers is one of the most common complaints voiced in the literature of the men's movement. All of this contrasts with the more traditional societies of the past where men plied their trades in and around the village or in the fields where the children could play and work alongside them on a daily basis. Young boys were gradually drawn into the regular world of men, imitating them and listening to their conversations and finally being initiated into manhood in some recognized manner.

One of the complaints of the woman's movement is that men avoid domestic responsibilities, leaving women with a disproportionate amount of responsibility for child care and domestic chores. And women point out that this patterns persists even while women themselves are more and more engaged in the workplace, so that they are asked to do double duty because men still typically avoid the "feminine" tasks of emotional engagement and child care.

Much of this behaviour on the part of men could be read as avoidance of responsibility, yet it could also be interpreted as an alienation from a broader and more authentic masculinity. Indeed, it seems clear that the separation of masculine roles from the domestic arena has served to exaggerate the natural differences of feminine and masculine characteristics, encouraging the over-development of certain aspects of the masculine character, while leaving others to languish. This is reflected in the stereotypical depiction of men as overly rational, aggressive, detached and lacking emotion, and women as emotional, passive, nurturing and compassionate.

Ironically, enforced specialization may have benefited men by encouraging them to develop capacities for coping in public life, such as competence with mechanical things, independent decision making, and responsibility for providing the material needs of the family and public good. Yet it has progressively isolated them from closeness to children, easily sustained emotional exchange with wives and extended family members, and spontaneous ease with that affectivity and intimacy which is an especially integral part of a full human life. With time men tended to become valued more for the efficiency and competence with which they carried out specific tasks, than by what kind of people they became, or how well they could relate to others. They were able to live lives in which there was much less scope for emotional and affective life than for pragmatic or functional activity. If they provided for the family in a financial way they were judged to be basically successful. It is this overdrawn specialization that is being reflected in much of the emotional emptiness and lack of meaning that some complain of today.

Early Masculine Development: A Problematic

The importance of early life experience for the individual has long been recognized in psychology, and the early period of the male child's experience in the family is no exception. Clearly there are other forms of cultural upbringing than the two parent family, yet this institution has had a widespread effect on the consciousness of many western men today, and recent feminist psychoanalytic thinkers have been providing some interesting reflections on its dynamics. As these theorists point out, during the modern industrialized era, the mother has become the primary care giver for infants and toddlers during the time the child forms its initial assessment of the world and begins to develop a sense of who it is.[4]

During this time it is primarily the mother who responds to the child's needs and keeps it in comfort. Carl Jung refers to this period as the Uroboric period, when the infant takes everything as one unified and ongoing mass of impressions, and experiences itself as intimately connected to all that surrounds it. Eventually, however, the infant begins to emerge from this state of merged consciousness into an awareness of being a distinct person. With the realization of its autonomy from the mother, the child vaguely realizes that, while it is free in a new way, it is also dependent on the mother and on significant others for its welfare. On the one hand it must please the mother in order to maintain the necessary relationship, and on the other hand it needs to assert its own need for uniqueness and individuality.

In his analysis of the hero myth, Jung points out that the mother, while essential and wonderful, can also be experienced as a restraining force which tries to hold back the emergence of independence. Her protective concern can be felt as suffocating possessiveness which wars with the child's unconscious need to discover and to assert its own independent self identity. And so in the great hero myths Jung sees the dragon as a symbol for the possessive mother who wants to devour the emerging independence of the child, while the "prized" princess to be rescued becomes the good mother, or the good feminine figure with whom the hero must eventually unite.

According to Chodorow and others, this normal growth of personal awareness, with its need for a healthy self image, poses particular

4 Chodorow, Nancy, *Feminism and Psychoanalytic Theory*, (New Haven: Yale University Press, 1989).

difficulties for the male child in comparison with those of the female child at this particular stage.[5] In contrast to young female children, when a young boy reaches the tentative awareness of himself as separate from his mother, he comes face to face with the fact that he is different in gender as well. It is nevertheless conveyed to him that he is different and is expected to be a little boy who is unlike his mother or sister. According to the neo-analytic interpretation, this means that he has to search for a new identity in a way that a little girl at this stage does not.

For the male infant, the mother whom he has primarily learned to identify with security and intimacy has withdrawn behind a veil of strangeness. This can register as a rejection or a betrayal of trust. Having learned intimacy with the mother only to discover that he is then to be distanced, he discovers that intimacy, while it is primarily learned from and found with women, can also be treacherous. One can be enticed and then abandoned.

For adult men in whom this transition was not negotiated though the availability of a reliably present male role model to whom he could turn for balance, there is the likelihood that women will still hold an aura of fascination and desire, but be experienced as dangerous and untrustworthy at the same time. Every time an adult woman begins to broach the defenses erected after his felt rejection as a child, the adult man becomes frightened of further rejection and discovers that he prefers to withdraw or at least to be reserved and self-protective. It is interesting to read how frequently successful, forward-looking men who consciously encourage their wives to start a new career after the children are older, shamefacedly discover that they are soon suffering deeply ambivalent feelings of loneliness and neglect when they have to come home to an empty house, or find their wives interested in things that no longer center around their husbands' lives or professions.

Similarly, there are many accounts attesting to the communication difficulties that arise between men and women where the woman finds the male partner to be emotionally reserved, perhaps interested in sex but not in a fully relational and sharing intimacy. Men are described by women as being incapable of conversing in the aimless way that both

[5] Chodorow, 1989; Baker-Miller, Jean, *Toward a New Psychology of Woman*, Boston: Beacon Press, 1976; Gilmore, David, *Mankind in the Making*, New Haven: Yale Univ. Press, 1990, pp. 9–29.

clarifies and promotes emotional bonding and communion. They tend to stress independence and activity, rather than the complex and seemingly purposeless rituals of interpersonal connection.

This frequent lack of easy facility with intimacy contributes heavily to the impoverishment of male confidence as well as posing a costly disappointment to women. Fortunately, however, we are not talking here of necessary consequences of child care, but the effect of exaggerated divisions between the roles of the genders in relating to children, conditions which are susceptible to change. The convergence of the absent father motif and the exaggerated separation of the social roles of women and men can coincide to exacerbate the young boy's struggle to find a clear direction for his identity when the mother becomes recognized as genetically different. Just at the time in his life when he cannot continue to model himself after his mother, a young boy is liable to be confronted with a weak or absent male role model in the home environment.

Where there is a warm and available male partner to whom the young boy can turn easily for reassurance, the discovery of the difference or "otherness" of the mother could be more easily negotiated and be an incentive to turn more directly to the father as the beginning of a long and mutually satisfying relationship. This obviously does occur in some cases, but it is also lacking in many instances. The emptiness and violence-prone demeanor of many young men, and the void-like feeling around the ideal of masculinity experienced by many adults, suggest that there is considerable difficulty in making and consolidating a successful transition to male role models and healthy masculine adulthood in our culture.

Today some men are taking a more active role in child rearing, yet this practice remains relatively rare, and the effective absence of the father still appears to be one of the perils of contemporary manhood. In fact today, the problem is increasingly effecting both parents.

Part III: Some Tentative Suggestions for a Positive Masculine Response: "Being Honest with the Real"

"Being honest with the real" is a phrase from Jon Sobrino encouraging us to look beneath the commonly accepted meaning of words and to deal with reality as it is rather than as we commonly think it to be.[6]

6 Sobrino, Jon, *Spirituality of Liberation: Toward Political Holiness*, New York:

His illustrative example is the phrase "the modern world," used as a description of the context within which most people live today. His point is that the image actually refers to a very small percentage of the world's population, those with access to wealth, technology, mobility, and freedom of choice. By misleading us into ignoring the truth that the vast majority of the world's people today inhabit a world of poverty, the phrase enables us to live with an image that is false and misleading.

The admonition to "seek the real" would be a useful paradigm in the search for a renewed vision of masculinity in the contemporary world. It could help to clarify the male situation as it is being lived today and promote a more fruitful cooperation with the struggle toward a more equitable style of relationship and social organization. There are a number of areas where it could be practiced.

1. To begin with, it would be good for men, as they reflect on their lives, to listen seriously to the experience of women. It is true that men are already exposed to a fair amount of more or less direct and assertive input from feminist interests in the public forum. But I suspect that it is rare for them to read serious feminist authors as they reflect on the impact of commonly accepted social mores and economic structures upon the daily lives of women, or on the image of women implied in many of the traditional concepts and structures. A brief perusal of Rosemary Ruether's discussion of the negative descriptions of women's nature in the writings of famous medieval theologians is an acutely embarrassing but educational experience.[7]

It is insufficient for men to acknowledge the need for greater fairness in the workplace, or even to be piously willing to understand the legitimate needs of women in general, without opening themselves to a genuine exposure to the thoughts and experiences of others who are different from them but who share the same space. In fact the insights of feminist authors can sometime's open men's eyes to realities and perspectives about themselves that they could never garner on their own.

In order to come to a full appropriation of masculinity as it is to be lived out in our times, men will have to understand better how they appear in the eyes of women, their frequent partners, companions and

---

MaryKnoll, 1988, pp. 14–17.

7 Ruether, Rosemary, "Anthropology: Humanity as Male and Female," in *Sexism and God Talk*, Boston: Beacon Press, 1983, pp. 93–115.

associates in life. At the same time however, it will ultimately be up to men themselves to arrive at a renewed and viable model of masculinity. They can not expect to grasp their own roots simply by piggybacking on the women's movement.

2. In the search for the real, men also need to realize and acknowledge the many dishonest advantages and insensitive attitudes that have become accepted as normal for them in many cultures. They need to look around the workplace, the churches, the malls, the home, and be alert to the presuppositions about who does what work, who gives most of the orders, makes most of the money, wields most of the influence and assumes most of the self importance. They need to be especially attentive to those quick explanations which instantly leap to mind to deny or deflect such criticism, and instead to subject these responses to the light of reality to see if they are not self-excusing rationalizations. To be "honest with the real" it often helps to put oneself in the other's place, even to the point of becoming indignant or embarrassed at the behaviour or attitudes of other men, or perhaps surprised by a novel perspective or appreciative insight into the uniqueness of maleness from a female point of view.

3. Sensitivity to the real world especially requires men to honestly face the nature of their own feelings in the context of their close relationships with women. They will need to notice where or in what way they experience ambivalence, self-protection, emotional reserve, fear of being open, and anxiety about being emotionally possessed. They will have to ask themselves who it is that is putting more into maintaining the relationship and how it is done. If they find themselves avoiding conversations about feelings and shared experiences, or being overly sensitive about being pressured or manipulated, it will be helpful to examine these feelings directly and to try and determine how objective they are, or if they are not simply unreflective fears. It will also be necessary to be open to discussing their relationships with significant others in their lives without attempting to resolve every issue as exclusively the problem of one party or the other. If they wish to understand themselves better, there will be a price, and this kind of reflectivity and interchange will be part of it.

4. It can be very rewarding for men to acknowledge the intrinsic value of their own feminine qualities as part of their search for maturity. Undoubtedly men need to accept their own desires for a certain

autonomy, but also to recognize their similarly strong needs for acceptance, love, gentleness and affection. They will have to refuse to concede that anything which is not self-contained and competent is effeminate or inappropriate, and acknowledge that they need people in their lives as much as they need work or achievement. In this regard too, it will be necessary for straight men to face and take responsibility for their often negative reactions to issues like gay rights or close relationships with other men. A good deal of masculine defensiveness and hostility may well derive from an outmoded and forced division between imagined feminine and masculine characteristics, as well as their ambivalence about their own identity due to an unresolved separation from their mothers and weak bonds with their fathers. Iron-clad, fearful and self-contained attitudes toward all signs of affection or intimacy with other men may well be a prominent symptom of the deeper pain and detached loneliness that brings much pain and unhappiness to the male psyche in general.

5. At the same time, in the current climate of critique and re-evaluation, men must learn to defend the validity of their own perspective and place in the world. They will have to discard unnecessary deference in the face of one-sided or ideological statements or expectations directed at them by popular feminism or particular women. It will be helpful to remember that women are engaged in the struggle to affirm their rightful place against entrenched privilege, and to appreciate that such struggles almost invariably produce exaggeration, narrowed focus, impatient and unrealistic expectations, and even manipulative methods. And men need to realize too that in many ways women are seeking to understand men and should not be presumed to have an unrestricted or privileged channel to male experience. There is much that women do not understand about typical male perspectives and experience. This is not to say that women do not have different and valuable things to convey, but it does mean that there is no magic finality about their impressions. They can only offer part of the data towards self-understanding for men, just as men for women.

6. Perhaps as the need for a mutual dialogue between men and women becomes more necessary and more possible, it would be helpful for both parties to nuance the concept of patriarchy, which sometimes sounds to male ears as an almost mantra-like explanation for any evil that has ever effected the earth, as well as a rebuke of everything mas-

culine. Thus, an awareness of some recent archeological research on ancient civilizations as it relates to the rise of patriarchy would be very useful for men, both for their own self-understanding as well as in their dialogue with women. It is important to realize that there have almost certainly been historical alternatives to the patriarchal or male dominated social structure which has now come to seem almost natural. Humanoids have been wandering the earth for an enormous expanse of time during which a great variety of social structures have come and gone before the relatively brief period of our own recorded history!

Experts speculate that in typical early gathering and hunting societies, probably consisting of bands of thirty or so people living off the land, women's relationship with children would most likely have been the strongest tie in the group, and their crucial power of giving birth and so sustaining the continuance of the clan or tribe would have invested them with an aura of mystery and power unknown to the modern technological mind. Indications of goddess worship in the artifacts and decorations honoring female fertility and birth are very common, and attest to a high status for women.[8]

It was only very recently in the life of the race that the beginning of plowed agriculture allowed the organization of large stable populations and complex civilizations. Then we begin to see written codifications of laws controlling ownership of property, inheritance, slavery, dowry prices and divorce regulations, class stratification of peoples according to power and status, warrior classes and the need for strong governing figures. It is in this process that the definite formation of patriarchal structures and male dominance gradually took hold.[9]

Thus it appears that patriarchy is a relatively recent development, and not the only social structure we have known. This is a heartening proposition for both men and women, for it proposes the possibility of flexibility rather than some hidden agenda of nature, and it should also be helpful in freeing men from the imputation of unreasonable guilt by reason of their gender. As many feminists would agree, hierarchy and

8 Ruether, Rosemary Radford. For a summary of the data supporting these developments see *Gaia & God: An Ecofeminist Theology of Earth Healing*. San Francisco: Harper, 1992, especially the chapter on "Paradise Lost and the Fall into Patriarchy," pp. 143–172.

9 Lerner, Gerda. *The Creation of Patriarchy*. N.Y.: Univ. Of Oxford Press, 1986.

privilege have often weighed heavily upon men as well as women, and in the long history of male/female relations there have been powerful examples of steadfast love, self-sacrificing care, or willing acceptance of responsibility running both ways, as a perusal of literature and famous biography would attest.

Rosemary Ruether has a provocative and timely reflection on the need for society to mitigate gender conflict by ensuring that adult males have a genuine importance and dignified role within a culture in order to balance the natural power of female affiliation with birth and the nurturing of children for the race. She notes the lingering tendency for conflict between the genders when men are held on the periphery of the great mysteries of birth and death and suffer a strongly accented sense of insecurity. From her observations of historical trends, she concludes that it is especially when there is little place for men in the life of the tribe that they have historically tended to turn their aggression against women. By contrast, in those tribal societies where sustenance has been readily available, and where there is no need to spend more than a third of the day providing for daily needs, there is a good deal of time spent by the whole group together at night telling stories and enacting bonding rituals, and just relaxing together. In these societies "men and women share fully the parenting of children from birth and the domestic work associated with daily life."[10] She suggests that today attention should be given to the tentative status of masculine roles in child care in the search for new accommodations among the genders. This raises again the perplexing problem of the rise of endemic unemployment in traditional male occupations.

7. Another area where men could "seek the real" is one that could be somewhat costly and acrimonious. In general men need to be more forthright in representing their own experience where it is being challenged or caricatured according to questionable ideological premises. Frequently in academic circles or public lectures, speakers of both genders will preface their remarks on the classifications of personality types by disclaiming their intent of placing people in set categories by these typologies, and then go on to leave men with such qualities as linear, cognitive thinking, aggression, individualistic, independence and power–oriented, and insensitive to nature: all the negatives of the age.

[10] Ruether, 1992, p. 171.

Women are granted intuition, creativity, relational skills and connectedness, affectivity, compassion, holistic concern for the environment, peacefulness concern for the global family: all the things which hold the promise of a more hopeful future for the planet.

In this context, it would be salutary for men to reflect more carefully on the foundations and nuances of their own experience, so that they can be articulate and explicit, instead of simply registering vague disagreement or silent irritation in the face of challenge. Anytime I have been present when some public disparity of this kind was assumed, I have heard irritation or disagreement from other men only after the event when only men were present. If dialogue is to move forward, men need to become more vocal when they feel that their own experience is being misrepresented. More fundamentally, they need to expand their self awareness and eventually to develop a legitimate and convincing symbolic base to represent the values and goals of an authentic masculinity for the future.

8. There is also a need for a "real" appraisal of the inner meaning of ancient masculine symbols. A positive first step toward a more solid sense of direction in the evolution of a new masculine self confidence could be a serious investigation of the inner meaning of some of the recurrent mythological symbols of masculinity as they have been expressed in the ancient legends and myths of the ages. The writings of Joseph Campbell on the Hero Myth or Eric Neumann on the evolutionary stages of human consciousness would be a helpful start.[11] Admittedly many of these symbols such as hero, king or warrior raise the spectre of much that people have come to find objectionable about our century, with its unrestrained competitiveness, violent militarism and brutalizing of humanity. Robert Bly's celebration of the Wildman in search of lost roots has come under criticism as being frivolous or trivial when compared to the conditions of women.[12]

Nevertheless, these ancient symbols have a very long history, and there are very few men, and perhaps women as well, whom these perennial themes can not move to deep and even thrilling responses when

11 Campbell, Joseph. *The Hero with a Thousand Faces*. N.Y.: Meridian Books, 1956) Neumann, Eric, *The Origins and History of Consciousness*, Princeton, N.J.: Princeton University Press, 1970.

12 Baumgaertner, Gill, "The New Masculinity or the Old Mystification?" *The Christian Century*, (May 29–June 5): 593–596.

they are handled well or expressed in an ennobling way in a great movie or drama. Qualities of strength, dedication, purposeful independence and goal seeking when dedicated to the service of the community still have a place if we are to defend our future and work together for a fair and ecologically viable world against the many forces of destruction that rage around us. It seems to me that ultimately we do have to acknowledge that there are two genders and that there are very probably differences between them which we will never succeed in eradicating. I do not believe that we will be able to simply begin anew and with perfect innocence to understand who men are today and tomorrow, without consulting the long evolution which is captured in these age-old myths. Contemporary authors such as Walter Ong with his *Fighting for Life; Contest, Sexuality and Consciousness*, as well as Patrick Arnold and more recently Dwight Judy represent contemporary attempts to adapt some of this ancient wisdom in creative ways.[13]

9. Perhaps one of the most delicate arenas for honesty on the part of Christian men will be that of life in the churches, particularly in the present instance, the Roman Catholic church. As an institution and a community dedicated to enabling the salvation of everything that is holy in the human character, there are many issues yet to be faced before we can become a model for the world. Here men need to ask if there is not something that we could learn from women about being true servants of the people of God. Men need to recognize who is responsible for sustaining much of the "informal" ministry in the church, and why that ministry is not recognized or funded more fully. They need to ask why only certain orders of ministry receive official recognition and remuneration, and why so much of the ministry typically done by women is expected to be voluntary. We need to ask ourselves seriously how we as men would feel if the use of exclusive language were reversed and how ready we would be to view it as a small matter.

Male religious might ask how we would have reacted if all of our required submissions on the renewal of our constitutions after Vatican II had to be submitted to committees whose voting members or supervisors were exclusively female. How would we have felt seeing our submis-

[13] Ong, Walter, *Fighting for Life: Contest, Sexuality and Consciousness*, Ithica, New York: Cornell Univ. Press, 1981; Arnold, Patrick, 1991); Dwight H. Judy, *Healing the Male Soul: Christianity and the Mythic Journey*, Crossroads, 1992.

sions frequently rejected or substantially altered, often in a manner foreign to our considered experience of life together? We need to ponder in our hearts the implications of restricting ordination, official leadership, and much of the authority to set moral and pastoral criteria in the church to one gender, and to pray carefully over the reasons for this restriction. And we need to ask ourselves how genuinely and deeply we actually desire an open and equal community in which each one's gifts are respected and made available for the service of all, and to acknowledge that our fear of change might be based on a grudging reluctance to surrender a reflected sense of prominence or influence. These exercises can lead us beyond simply arguing within ourselves or with others about the need or advisability of changes, but could help us to reach a more contemplative and honest appraisal of our own hearts and of the world that we most genuinely wish to cooperate in co-creating.

10. Finally, perhaps both men and women need to remember the wonder and the natural intensity that frequently exists in the relationships between women and men. When a person of one gender walks into a room where there are only members of the opposite sex, there is usually a certain new charge added to the group. And this difference is always part of our consciousness, whether we are young or older, married or single, celibate or seeking. Perhaps this ambience is even heightened but confused today with the increased sensitivity to harassment or inappropriate behaviour. This attraction and excitement will always remain part of human experience and it continues to produce a great deal of the spice that we find in life. Much human joy and inspiration has come though its auspices. The lure of romance fills our literature, drama, magazines, newscasts, dramas, movies, myths and history. It produces courtship, delight, betrayal, disappointment, ecstasy, longing, yearning, grieving; it is filled with promise, fascination, danger. A person gains and loses, is accepted and rejected, finds companionship and loses carefree independence. Old memories of youthful romance fill the background of our imaginations and only strike through our forgetfulness on rare occasions, but they can be incredibly life-giving and humbling when they do. The time we threw a rose into the book-bag of a surprised girl and ran off stunned with embarrassment, or boasted with the other boys about some heroic escapade but let it die as soon as we were aware that the girls were not listening, or the first time we really felt the stilling wonder of what love for another person could mean.

So if we as men are groping to find ourselves as men today, we are going to do so in the context of the ancient dance. Human beings will be unable eventually to form a new life and society without being men and women together. Whether we prefer it or not, we are destined to live in each other's company, as are men with men and women with women.

Seasons *of a* Woman's Life, Seasons *of a* Woman's Faith

Patricia Cooney Hathaway

A few years ago I received an invitation to conduct a day of reflection on women's spiritual journey. I chose the title, "Seasons of a Woman's Life, Seasons of a Woman's Faith," for two reasons. First, I wanted to explore the interweaving of the human and spiritual journey in women's lives. Second, I wanted to see if the theory of adult development described in Daniel Levinson's book, *Seasons of a Man's Life*, could be applied to women. His subsequent book, *Seasons of a Woman's Life*, published in 1996, confirmed my original conclusions.

The women who attended this day found the description of the seasons of their lives very helpful. It enabled them to name their experience and, more importantly, to see how God works through everyday life to draw them into a deeper relationship. Since that time, I have received numerous invitations from women's groups all over Michigan to share this material with them. As a part of the day, I ask the women to describe the meaning of their spirituality at this particular time of their lives. I have included some of those descriptions in this work as a powerful illustration of the interweaving of the human and spiritual journey. This essay, then, describes the seasons of a woman's life and how women learn to recognize and respond to God in the concrete circumstances of their lives.

The Season of Spring: 17–28

The early adult transition, ages 17–22, provides a young woman with the opportunity to imagine various possibilities of her self and the world in the future. In this initial period of the early adult era, a crucial task is the forming and living out of a *Dream*. The Dream refers to a sense of self in the adult world. It contains the quality of a vision, an imagined possibility that generates excitement and a zest for life.

In religious language, one's dream refers to one's vocation. Since Vatican Council II, the Catholic understanding of vocation has undergone a dramatic transformation. In the piety of past generations, to "have a vocation" meant to be a vowed religious or priest. The rest of us, married or single, muddled along without any special religious identity or sense of calling. Today, a more adequate theology of Baptism insists that every adult Christian has a vocation. God invites each of us to give our life's energy to a particular way of loving, working and contributing to the bringing about God's kingdom in the world.

Developmental psychologists tell us that those who build a life structure around a dream in early adulthood have a better chance for personal fulfillment. Those who betray their dream in their twenties will have to deal later with the consequences. Those who have no dream of their own and piggy-back on someone else's will inevitably drift into an early life structure with no particular sense of self in the world. Levinson's study identifies the variety of choices made by women in the early adult transition:[1]

1) The Relational Dream: forming relationships, most typically through marriage and motherhood. Ideally, a woman seeks a relationship with a special man and forms an initial life structure with him. Her dream is to become a certain kind of wife and mother. Out of the forty-five women interviewed, twenty-five women chose marriage in the early adult transition, eleven of whom held outside jobs only when financially necessary.

2) The Individualistic Dream: becoming a certain kind of individual through forming a career identity. Levinson identifies these women as attempting in early adulthood to actualize what he refers to as the "internal Anti-Traditional figure." For many, the dream was not highly formed in the early twenties, nor did they have a clear vision of a long-term career.

3) Women without a Dream: Levinson observed that many of the women had a moderate to severe crisis during the course of the early adult transition. They had a basic sense of being overwhelmed, of having no way to form a minimally good enough life as an adult. For some the crisis involved considerable turmoil and conscious suffering, others were not clearly aware of the extent of their difficulties.

4) Women who chose Relational and Individualistic Dreams.

A fourth group of women chose to balance relational and individualistic dreams at the same time. These women wanted occupations which

[1] Daniel Levinson, *Seasons of a Woman's Life*. New York: Knopf, 1978.

fulfilled them as individuals. They wanted to combine career with the relational dream of husband and children.

The essential task, then, for the spring of life is the forming and giving shape to one's dream—one's sense of self in the adult world. Often it is provisional at best. Yet, the form that the dream takes provides the first adult structure for entering the adult world.

Interpersonal Faith

What is the invitation for our spiritual life at this time? It is the discovery of faith as a personal relationship with God. In the spring of one's life, the ability to fall in love on a human level often functions as a moment of grace (Kairos) on a spiritual level. It involves the discovery of God as *personal* in a way that is possible only when one has grown into the capacity to fall in love with another. In other words, it is the experience of romantic love as *passionate* in our human experience that foreshadows what falling in love with God can be like. Now we are capable of experiencing God as a divinely personal Other—friend, lover, companion—who invites us into a personal relationship.

In the Old Testament literature, the Canticle of Canticles expresses this experience. In the New Testament, the stories of discipleship are not initially stories about mission, but about relationship with the Lord. The disciples became attached to the Person of Jesus to such an extent that they gradually came to share his vision of life and ministry and chose to live as he lived and carry out his mission. The following descriptions of faith illustrate the spiritual awakening that can take place at this time.

Young adult woman: age 19

> In high school I experienced a deep desire for intimate personal relationships. I believe my exploding sexuality and adolescence was a major factor in this. My relationship with God became focused on a personal relationship with Jesus.

Young adult women: age 21

> As I began to gain more perspective on myself and my relationship with men, so also could I begin to detect a new sort of hunger emerging for exploring more deeply my relationship with God. It was as if, through this crisis that I confronted, I had discovered a new, more personal and compassionate God—in many ways a God far different from the distant and demanding father God of my childhood.

These faith experiences are *possibilities* not *actualities*. There are a variety of faith responses that occur at this time. Many young adults continue to live the "faith of the clan." In other words, what was good enough for mom and dad is good enough for them. For the most part, faith is taken for granted.

For many others, however, adolescence and young adulthood begin a time of questioning, and often, a time of drifting away from their religious faith. In her book, *The Birth of the Living God*, Dr. Ana-Maria Rizzuto, provides insight into what is happening at this time.[2] The major thesis of her work asserts that one's God-image needs to be recreated in each developmental phase of life if it is to be found relevant for lasting belief. Most of us do this unconsciously. A potential problem occurs, however, when the initial God image (our image of God as children) remains untouched as we continually revise our parental images and our self images during the developmental periods of the life cycle.

Dr. Rizzuto's research indicates that it is during late adolescence and early adulthood that most people who cease to believe drop their God. When one's God image does not keep pace with the changes in one's self image, it is soon experienced as ridiculous, irrelevant, threatening or dangerous. Belief ceases because it loses meaning or remains developmentally immature. The possibility of discovering religious faith as a personal relationship with God begins here. It is often accompanied by high energy and lofty ideals which are concretized in the desire to do something "beautiful for God" in the transformation of the world.

[2] Dr. Ana Maria Rizzuto, M.D. *The Birth of the Living God. A Psychoanalytic Study.* Chicago: University of Chicago Press, 1979.

The Summer of a Woman's Life: 28–39

The issues and tasks of the summer of a woman's life are contingent upon her satisfaction with the life structure formed during the 20's. One impetus for change in this period is that the entry life structure is necessarily flawed to some degree. During the early adult transition women are on the threshold of a new world and a new self, but they often know too little about either to have much wisdom in making the choices that bring them into adulthood. Levinson's research resulted in the following observation about women in their thirties.

> In the Age 30 transition, many of the career women, like the homemakers, often came to feel that they had gone through the twenties with minimal awareness of themselves, of their future lives, and of their relationships with significant persons and groups.[3]

The age thirty transition, then, which typically begins at age twenty-eight and ends at age thirty-three, is a time for second thoughts. What have I done with my life thus far? What do I want to make of it? What parts do I want to give up or appreciably change? Are there new directions to pursue?

For women who chose marriage and family in their twenties and who are happy with their choice, the age thirty transition is barely noticeable. For women who are dissatisfied with their choice, this is time for reassessment. What needs to be done to enrich family life? Can I stay in this marriage? Do we now want to have children? Do we have enough children?

For women who chose the individualistic dream, the often harsh reality of what it takes to 'make it' in the world of work brings on a time of reappraisal. Levinson's research concluded,

> Many career women went through a marked, often painful process as they became aware of the competitive struggles, the "politics" of organizational life, and the diverse obstacles to advancement for women. The work world was not as caring nor as rational as the young women had expected, and the career woman's progress within it depended upon much more than their own ability and expertise. The career women had to reappraise their occupational aspira-

[3] Levinsin. *Seasons of a Woman's Life.* New York: Knopf, 1996, 303.

tions and the relative value they placed on work, family and other aspects of life.[4]

For career women who were also single, the internal pressure to stabilize their lives with marriage and family became more paramount. Research data confirms that at this time of reassessment, women must come to terms with the issues of marriage and parenting in both a quantitatively and qualitatively different way than is the case for men. Even though today more women put off having children until later, reaching the age of 35 raises the following questions: What can I do to make relationships more central? Will I ever have children? Can I be happy just as a career woman?

For the women who endeavored to integrate marriage, family and career, the age thirty transition is a time to reassess the quality and progression of both dreams and their ability to achieve a balance between the two.

Reflective Faith

A helpful term which describes the faith response of many women in this season of life is *Reflective Faith*. Reflection is the act of deliberately slowing down our habitual processes of interpreting our lives to take a closer look at our experience and at our framework for interpretation. We reflect when something happens to us that we cannot readily fit into the interpretive categories we normally use to make meaning in our lives. In relation to our faith life, there follows a movement from an unreflective to a critical owning (or rejecting) of religious faith as "*my* faith."

This awakening can happen through a single event or can occur over a period of time. It can stem from times of difficulty and crisis: a decision over birth control or abortion, divorce, the death of parent, spouse, or child. It can be brought on by the experience of personal harassment, oppression, discrimination at the work place or church. Such crises raise questions such as: Why did God do this to me, to my child? How will I survive?

On the other hand, the awakening can happen through a positive life experience: falling in love, giving birth to a child, or experiencing

4 Levinson, *Seasons of a Woman's Life*, 297.

the unconditional love of a spouse over a long period of time. Such experiences, negative or positive, serve to bring the question of meaning, and therefore, the question of God, to the conscious surface of one's life. It leads women to look seriously at the image of God inherited from family and church, and to ask: Who is God *for me?* Where is God for me? How do I experience God in my life? The living with and gradual resolution of these questions often leads women to a new image of God, self, and the role of the institutional Church in their lives. One woman writes:

> I had just turned 30, my husband, 32 when we gave birth to our sixth child. I was raised Catholic. My husband wasn't a religious person, although he was a very good man. At my insistence we followed natural family planning. Obviously, it wasn't working. We had to do something. I didn't want to use any artificial birth control, because my church considered it a serious sin. My husband came home one night and told me the issue was settled—he had had a vasectomy. I thought he would go to hell. He told me that he couldn't believe in the God I believed in. His God understood the need for people to make difficult decisions. What followed for me was a long, painful process of re-evaluating my whole faith—a faith that I had accepted unquestioningly from my Mom and Dad.

Wrestling with these issues has led many women, although certainly not all women, to the controversial issue of the Women's Movement and Feminism.[5] One woman from Levinson's study remarks:

> At 27 I applied to business school and began at 28. I started business school with no clear idea where it might lead. I was not "liberated" and wasn't aware of the Women's Movement until I got to business school. Then it hit you like a sledgehammer, the *need* for a Woman's Movement. I'd never been particularly aware of discrimination, but there were 28 women in my class of 500, and you really felt a minority... The most blatant example was the marketing pro-

[5] The meaning brought to Feminism in this study is a self-conscious movement that opposes any ideology, belief, attitude or behavior—either in society or in the church— that establishes or reinforces discrimination on the basis of gender. Its ultimate aim is the liberation of all women and men who suffer oppression of any kind and equality among all persons regardless of gender.

> fessors. They made their resistance to women quite obvious. They'd always call on me to give comments on the "consumer's point of view"! So I went from being totally unaware to being militant in about three months (laughs.)[6]

When the experience of discrimination or oppression, be it subtle or overt, is experienced within the Church, it can have the same effect as being hit over the head with a sledgehammer, causing one's whole faith-life to come into question. After completing a Master's degree in Theology and Pastoral Ministry, one middle-aged woman shares her experience when applying for a position in parish ministry.

> Having been groomed for a position belonging to a friar, I was devastated when the priest denied me the position because I was a woman. I entered a state of numbness.... It lead me to question everything. Can I and can my daughters realize our vocation as baptized Christians within the tradition of the Catholic Church?

For another woman, it was hearing the hymn, *Faith of our Fathers.* "I have heard that hymn a hundred times," she said. "But all of a sudden it hit me. What about the faith of *our mothers?*" For another it was the priest refusing to allow her daughter to be an altar server. For other Catholic women, their particular questions are equally unsettling: "If I can no longer pray to God as father, whom do I pray to?" "Does the very fact that I have these questions and feelings mean that I am a feminist?"

Women's struggle with these questions doesn't mean that they are necessarily identifying themselves with Christian Feminism. The more important issue is how living with these questions influences women's faith life and spirituality.

In the summer of a woman's life, then, something happens. It may be a moment of crisis or it may be nothing more than the gradual process of maturing. Whatever the experience, it causes a woman to look at her faith reflectively and ask the questions: What does it mean *to me?* Who is God for me? One's faith is no longer the faith of mother/dad, husband, or minister. It is *Mine.*

6 Levinson, *Seasons of a Woman's Life*, 274.

The Fall of a Woman's Life

The season of fall in a woman's life begins with the mid-life transition which occurs between the ages forty to forty-five. The term was coined by the psychiatrist, Carl Jung, to describe a serious and very real life experience.

Jung maintained that the tasks of the first and second halves of life are different. The first half of life is oriented primarily to adaptation and conformity to the outer world. The second half of life is oriented primarily to adaptation to the inner world.

Jung termed the development of the person during the first half of life "ego development." The ego is the conscious personality. In Jung's psychology, the whole personality, or the Self, is made up of both conscious and unconscious elements. The task of the second half of life is the gradual integration of the conscious and unconscious dimensions of the personality in the realization of the true Self. Jung referred to this process as Individuation.[7]

Jung stressed the significance of the second half of life, for it is during this time that the neglected, repressed, or unattended dimensions of our personality begin to clamor for attention and integration. Jung identifies these as "voices from the other rooms." He cautioned women and men to listen to the neglected voices of the inner self and to make the changes necessary to integrate these dimensions of the personality into their conscious lives, or middle adulthood would be much more difficult.

While there are many transitions in life, the mid-life one is the most critical. At this time one's perspective changes from "there is still so much time ahead," to "how much time do I have left." If a woman is to make fundamental changes, she must make them now before it is too late.

Research on the seasons of a woman's life indicate that the central issue for women at mid-life is personal integrity; that is, becoming one's own person. The socialization process of women in church and society has made the issue of personal autonomy a very difficult one for women. Women are socialized to live through others in a way others never live through them. The task of mid-life and beyond for women, then, is to take responsibility for the development of those talents and qualities in a way that will allow for the emergence of the whole self.

[7] C.G. Jung. *Modern Man in Search of a Soul.* New York: Harcourt Brace, 1933.

Mid-life Transition or Crisis

Levinson's research draws a valuable distinction between a mid-life *transition* and a mid-life *crisis*. The mid-life transition functions as a bridge between early and middle adulthood. That is, the woman's life-structure remains intact with minor or major modifications to accommodate new goals. A thirty-nine year old Episcopalian woman recalls her effort to fulfill a life-long dream.

> I was thirty-nine years old, washing dishes at the kitchen sink when it hit me—it's either now or never. Ever since I was a little girl, I wanted to be a minister. I realized that if I didn't begin Divinity school soon, it would be too late. Within a few days I called my husband and teenage children together and told them my plan. They were not happy about it. But I'm determined to see it through. I believe I have the calling and the talent to be a good priest.

A mid-life crisis occurs when the whole life structure one has built over the years is 'up for grabs.' A woman whose divorce at age 50 ended 29 years of marriage states:

> "I was looking forward to having my companion to grow old with. After raising four children, I felt it was all down hill from now on. We could enjoy the profits of all those years of effort. Instead I found out my husband had a different agenda. My personal feeling is my husband didn't want to come face to face with his mortality. He didn't want to be a grandfather."

Be it mid-life transition or mid-life crisis, a woman eventually must get on with re-prioritizing her present life structure or reestablishing a new one to deal with the tasks that begin at mid-life and continue through the middle years.

Levinson's research identified four primary tasks during middle age. Each task involves the reintegration of a fundamental polarity of life: Young/Old, Destruction/Creation, Masculine/Feminine, Engagement/Separation. Each of these pairs forms a polarity in the sense that the two terms represent opposing tendencies or conditions. Superficially, it would appear that a person has to be one or the other and cannot be both. In actuality, both sides of each polarity coexist within every self. For example, we are both young and old at each stage of development.

The polarities present problems of conflict and integration through the middle years. How a woman prepares for and resolves these polarities determines the meaning and gracefulness of her middle and late adult years.

Season of Faith: Paradoxical Faith

The term which describes the quality of faith which many women live out of at this time is *paradoxical* faith: how can new life come out of death? Some of the most helpful descriptions of what this faith looks like comes from contemporary reflection on the spirituality of John of the Cross.

In an article entitled, "Impasse and the Dark Night," Sister Constance Fitzgerald explores the experience of limit, void, and emptiness that describe many women's lives and their faith at various times during middle adulthood.[8] She points out that we all begin ministry, marriage and/or family life with great desires, and we may spend many years experiencing the satisfaction of those desires. This is how life should be. Yet, the time will come in our development when every significant God relationship, human love relationship, marriage or ministry will reach an impasse; that is, we will experience its limits, its inability to satisfy us in the way it did before.

Impasse experiences are those situations that choose us, rather than being chosen by us. They are life-situations in which there is no way out, no way round, no rational escape. They may be crisis situations such as divorce, or the loss of husband or child due to terminal illness. Or they may be normal life passages such as looking in the mirror and seeing unmistakable signs that the aging process has begun and that there is little one can do about it. To make matters worse, God doesn't seem present. It is a time when we experience the limits of self, others, our vocation, God.

When such situations are put within the interpretative framework of John of the Cross's Dark Night, the person is re-assured and energized to live. John of the Cross spoke of the dark night to show what kind of affective education is carried on by the Holy Spirit over a life-

[8] Sr. Constance Fitzgerald, "Impasse and the Dark Night," *Women's Spirituality.* Edited by JoAnn Wolski Conn, New York: Paulist Press, 1988.

time. He articulates what we all experience: there is a dark side to human desire. We can be possessive, selfish, ego-centered. Consequently, John delineates the work of the Holy Spirit who moves us from a desire or love that is possessive, entangled, complex, selfish and unfree to a desire that is fulfilled in union with Jesus Christ and others.[9]

John points out that in the process of life our affectivity is not suppressed or destroyed, but gradually purified and transformed. The progressive transformation of the human person takes place within our every day life experience *through* what we cherish or desire, that which gives us security and support, where we are most deeply involved and committed, and in what we love and care for most. The experience of limit, void or emptiness does not necessarily mean there is something wrong with our commitment or our prayer. Rather, it is a sign we are undergoing a time of purification and maturation in our marriage, ministry or relationship with God.

The problem we face today, however, is that men and women in first world cultures are not educated for darkness. Every commercial, every advertisement we see promotes instantaneous gratification. Consequently, we are tempted to look for external ways to replenish the interior wasteland: frenetic activity, leaving ministry for marriage, addiction to drugs, alcohol, work, shopping, or T.V., depression, or withdrawal. It is precisely in the throes of such crises that people abandon God and prayer, a marriage, a friend, a ministry, a community or a church, forfeiting the possibility of new vision, the maturity of love and loyalty, dedication and mutuality that is on the other side of the dark night. The challenge is to learn how to how to stay with the darkness, to let it work out its purification and maturation in us. For it is the darkness that holds the promise of a new life, new vision and experience of God, self, and others. A forty-three year old woman, living with the aftermath of a difficult divorce writes:

> I can absolutely attest to the fact that whatever issues are not recognized, confronted, accepted, and resolved in the various stages of human development, will emphatically make themselves known and demand for clarification at whatever cost to the individual. In my spiritual journey at this time of my life, I am in the clearing in the woods. I look back at the forest I have just come out of, that

[9] Ibid., 102.

> dark, scary forest, filled with the "demons" and the "strange noises" from a myriad of painful experiences beginning with my early childhood, and I wonder how I ever found my way out of it. But of course I know now, that the God I was desperately searching for and who seemed so distant and non-existent was leading the way.

The tragedy is that many people refuse to make this passage. They are brought time and again to this point, yet they continue to turn away from it and refuse the call to deeper maturity over and over again in our lives. The forty-three year divorced woman describes the promise of new life on the other side of the dark night.

> So now I am in the clearing; I am catching my breath so to speak, resting in the quiet, listening. I am discovering the Jesus who led me through the woods and I am discovering my real self. I don't think Jesus could work through my "false" self—I couldn't work with myself! But I now know that He loves me and those times I cried out, "Help" and thought I was hopelessly lost in the woods were the times He took my hand and said, "This is the way out." My resting in this quiet, sunlit place is not so much a reward or consolation as it is a gathering of strength, of becoming rooted in my faith and in Jesus.

The years following this dark night usually find women at their best. There is an experience of integration. A woman goes about life with more peace, freedom of spirit, inner delight. She is successful in her life project, career, family life, ministry. She may have long peaceful years in a community or a marriage. She experiences a mature kind of mutuality in relationships—a kind of assurance that comes from being loved. There is a freedom from structure that allows her to live within the structures of her life with freedom. She experiences a deeper sense of and trust in God's abiding love for her. She may feel compelled to share God's love with others through some type of Christian ministry. Often she has become what the Christian tradition identifies as a contemplative-in-action.

The Season of Winter: 66+

In her book, *Winter Grace, A Spirituality for the Later Years*, Kathleen Fischer claims that by the year 2000 there will be over 35 million persons in the U.S. 65 or older. Estimates are that by the year 2050, that number will double. This phenomenon of the large aging population is clearly one of the most significant developments of our century.[10]

Thanks to research in the field of aging, we now better understand many of the physical, psychological, social and economic aspects of the aging process. But we have not yet fully answered the deeper questions emerging from our longer life expectancy. Does anyone know *how* to live the last years meaningfully and joyfully?

The fact is that aging is both descent and ascent, loss and gain. Certainly one of the most obvious yet difficult experiences the elderly deal with is precisely the reality of loss. Loss of husband or wife, brothers and sisters, friends, good looks, financial resources, loss of control over one's own life. One woman describes a four year grieving process during which she lost two brothers and one sister to cancer. "I have a huge hole in my heart. In the space of a couple years, all my brothers and my sister are gone. I feel so alone."

A second real concern for the elderly is becoming dependent upon their children. This kind of dependence takes many forms. A woman falls down and breaks her hips. She had lived alone but now she must move in with her daughter. "I feel like such a burden. I try to stay out of the way as much as I can."

In the face of such difficulties, what are the gains of this time of life? In her poem, "The Contemplation of Wisdom," May Sarton suggests that life's many losses yield the gift of wisdom.

> Partaking wisdom, I have been given
> The sum of many difficult acts of grace,
> A vital fervor disciplined to patience.
> This cup holds grief and balm in equal measure,
> Light, darkness. Who drinks from it must change.
> Yet I am lavish with riches made from loss.[11]

This wisdom often expresses itself in gratitude for and enjoyment of the simple pleasures of everyday life. One woman reflects that during much of our life, we measure time by its duration. We rush through one task

10 Kathhleen Fischer, *Winter Grace*. Spirituality for the Later Years. New York: Paulist, 1.

11 May Sarton,*Collected Poems 1930–1973*. New York: W.W. Norton and Company, 1974, 409–410.

in order to hasten on to the next. As we reach late life, this preoccupation with quantity can give way to an appreciation of the quality of time and the desire to use well whatever time one has left.

Trustful Faith

The quality of faith needed to guide the elderly through this season of faith is a trustful faith. It is a faith which believes that God will be true to her promises to be with us in all things—ascent and decline, gain and loss, success and failure.

Women in the winter of life experience the joy and satisfaction that comes from looking back on former seasons of life and realizing that through God's guidance and fidelity they have negotiated the seasons well. They continue to live this season with enthusiasm and vitality.

Yet other women experience a need to trust that God continues to be with them through difficult times. One seventy-year old wife, mother and grandmother writes:

> I think that trust is a good word to describe it. But I was a *long* time coming to that. My faith and my personality were *very* immature until past 45. You described the "happy housewife" who suddenly reached a mid-life crisis and found that she wanted a career. I was the reverse—an "unhappy housewife," hating scrubbing floors and cooking and wanting a career. Then suddenly I found that was not really what I wanted. And I began to grow.

The most difficult challenge of the winter years is finding a way to deal gracefully with the decline, losses and the inevitability of death. Sister Constance Fitzgerald offers the interpretative framework of John of the Cross' Dark night of the Spirit to find meaning in the experiences of loss. In a series of tapes on "The Meaning and Recognition of the Dark Night Experience," she describes how the dark night of the spirit in its deeper dimensions is a metaphor for death and in the light of death, for the aging process.[12] Physical death is the last and ultimate experience of the dark night. It is a metaphor which sheds light on the various experiences of darkness and trauma that actually foreshadow this final

12 Sister Constance Fitzgerald, OCD, "The Meaning and Recognition of the Dark Night Experience," Alba House Cassettes, 1991.

purification that most of us cannot evade. But why does it have to be this way? John of the Cross explains:

> This dark night is the inflow of God into the soul which purges it of its habitual ignorance and which contemplatives call infused contemplation. Through this contemplation, God teaches the soul secretly, in darkness, and instructs it in the perfection of love without its doing anything, or understanding how this is happening.[13]

In describing how this inflow of God takes place, John of the Cross talks repeatedly about the powers of the human person: intellect, memory and will. The purification centers on them in this dark night. It is different from the purification of our affectivity in the dark night of the senses which was described in the season of paradoxical faith. John maintains,

> God divests the faculties, the affections,
> the senses both spiritual and sensory, interior
> and exterior. He leaves the intellect in darkness,
> the will in aridity and the memory in emptiness and the
> affections in supreme affliction.[14]

When John says the intellect is emptied and left in darkness he is warning us that the time will come when our carefully constructed meanings and philosophy of life will fall apart before our eyes. Everything we have accumulated intellectually through our education and life experience; everything that has given us security, faith, God, loses its significance. The mind that is very full on one level—the level of a lifetime of knowledge and experience—is in total darkness on another level—the level of meaning. Fitzgerald points out that people become very quiet at this time. It is very hard for them to talk about what is going on inside them. This is the blackest time of the night when the last delicate shading of the image of Christ is being etched in—the image of Christ crucified and abandoned.

In describing the purification of the memory, John speaks about a time when the imagination can no longer connect life's memories to pro-

13 John of the Cross, *The Dark Night.* in *Collected Works of St. John of the Cross.* trans. Keiran Davanaugh, O.C.D. Washington, D.C. Institute of Carmelite Studies, 1979.

14 Fitzgerald, "Dark Night Experience," Cassette #4.

duce meaning and hope. One can speak of emptiness in the memory not because one remembers nothing, but because all that the memory holds now seems an illusion and a mockery. Here, self-knowledge is acute and we remember more our failure than our successes. John describes this experience.

> A person feels so unclean and wretched that it seems that God is against her and she is against God. There is a deep immersion of the mind and memory in the knowledge and feelings of one's miseries and evils.[15]

Abandonment and the seeming betrayal of one's trust is the hallmark of the dark night of the will. However it happens in one's life, what or who one wants and needs and clings to more than anything in life, what one cherishes above all else, is cut off, taken away, withdrawn.

A concrete illustration of this experience is found in the decision to place a parent in a nursing home. So often an immediate deterioration of the body and mind sets in. While it may be a necessity, the parent often feels abandoned and betrayed by the partner and children he or she cherishes and counts on for the assurance of meaning through the final passage of life.

What is the purification and transformation that is taking place here? In stanza three of the *Living Flame of Love*, John describes our human powers of knowing, willing and remembering as deep caverns of feeling that have a capacity for the plenitude of God. As long as these deep caverns are full of other things, we can't genuinely ache for them to be filled with God. Only when we experience the fragility and breakdown of what or whom we have actually stacked our lives on, and experienced the failure of our life project and the shattering of our dreams and meaning, only then do we thirst and hunger and yearn for these caverns to be filled with God. Emptiness is what effects the exchange.[16]

What is John's response to this dark night experience? There is only one option which throws open these caverns to God—the theological virtues, faith, hope and love. We now must have faith in God for God's sake, hope in God for God's sake, love of God for God's sake. These virtues are activated solely by God's dark presence. And they come into their full flowering at this level of maturity. Yet at this stage we do not

15 Ibid, Cassette #4.

16 Ibid, Cassette #4.

feel faith or hope or love. This posture, however, must be maintained in our prayer and in all the ramifications of our life and our relationships. John reassures us:

> This night, this darkness guided me
> more surely than the light of noon
> to where he waited for me,
> him I knew so well, in a place
> where no one else appeared.[17]

There is a transformation of the whole person and the total redirection of energy for God in the world. The person no longer knows or understands with the vigor of her own natural light, but with the divine light, with the perspective of God. Through this dark night of the spirit the intellect, will and memory are transformed so that we learn to know, to love and to remember from God's perspective. John of the Cross describes the goal of this dark night.

> How gently and lovingly
> You wake in my heart
> Where in secret you dwell alone;
> And in Your sweet breathing,
> Filled with good and glory,
> How tenderly You swell my heart
> with love. [18]

While gradual decline of old age or terminal illness are often the context within which this dark night of the spirit occurs, they are not the only life experiences to do so. Divorce, illness or death of a child, loss of job, struggle with alcoholism and its destructive effects on a family, are examples of other life experiences that can catapult one into this deeper experience of purification and transformation. People who come through this experience and live out of a new sense of God's presence and God's vision are powerful forces of conversion and reconciliation for others.

17 John of the Cross, *The Dark Night.* in *Collected Works of St. John of the Cross.*

18 John of the Cross, *The Living Flame of Love.* in *Collected Works of John of the Cross.* Trans. Keiran Kavanaugh O.C.D. Washington, D.C. Institute of Carmelite Studies, 1979.

Conclusion

A fundamental Christian truth that has weaved its way through this essay is that we meet God within the people and events of our ordinary life experiences. The challenges, crises, tasks, successes and failures of each stage of life can be invitations to a deeper relationship with God if we have the eyes to see. As an illustration of how this interweaving of our human and spiritual journey applies to life, let me close with the reflection of a woman on a passage from the gospel of Luke as she finishes a day of recollection on this theme.

> Leaving the synagogue Jesus went to Simon's house. Now Simon's mother-in-law was suffering from a high fever and they asked him to do something for her. Leaning over her, he rebuked the fever and it left her. And she immediately got up and began to wait on them. (Luke 4: 38–39, *Jerusalem Bible*)
>
> Jesus leans into the woman's space to heal her. Well again, she rises and serves him in the way she knows best. As an older woman she meets his needs for food and refreshment with her particular expertise...that of a homemaker.
>
> After finishing this retreat, my prayer is that Jesus will enter my personal space to heal those darkened areas within my spirit that weaken and incapacitate me so that I may serve in ways that will bring myself and others to the fullness of union with Him.

This woman expresses my hopes and desires for every woman reflecting upon the seasons of a woman's life and seasons of a woman's faith. May Jesus enter our personal space to heal those darkened areas that weaken and incapacitate us, so that we may choose life with greater clarity and self-awareness, and in the process, lead others to do the same.

Spiritual Darkness *in* Women's Lives

Joann Wolski Conn

Women's experience of spiritual darkness is best understood, I believe, by attending to *experience*, that is, to this darkness in all its particularity and concreteness. Therefore, I would like to focus on particular female roles and the concrete lives of two women: Jane de Chantal and Cornelia Connelly. My questions are these. First, how did their roles as women in their families, the Catholic Church, and society influence their experience of spiritual darkness, that is, of emptiness, of struggle to adhere to God when divine presence seemed absent or far away? Second, how does their experience modify and contribute to the dominant tradition regarding spiritual development?

Jane's and Cornelia's experience of being "in the dark" regarding God's presence and desires for them is radically shaped by their roles as wives, mothers, widows and foundresses of religious communities. These women's lives are so fascinating, and their experience of emptiness so profound and poignantly described in their own words, that I am frustrated by my inability to convey it with any adequacy in a few pages. Nevertheless, I will summarize key aspects of their stories and notice how being a *woman* shaped their experience of spiritual darkness.

Jane de Chantal (1572–1641)

Daughter of the president of the Dijon (France) parliament, Jane's family arranged her marriage to a man whom she grew to love deeply and with whom she had four children who lived beyond infancy. After eight years of marriage, her husband was accidentally killed by a friend in a hunting accident. This marked a dramatic beginning of Jane's process of spiritual darkness, opening her deepest self to greater love, self-knowledge and movement beyond her current ego-boundaries. Her father-in-law then insisted that Jane, newly widowed, move into his household and manage his estate as she had previously managed her husband's otherwise he would disinherit her children. He needed to threaten such dire consequences for her young son and three daughters because

Jane cringed at moving into his house where his housekeeper was also his mistress, mother of their five illegitimate unruly children, and domineering arbiter of final decisions. Jane's growing desire for a deeper relationship to God became a painful yearning as she strove to raise her children in this setting, fulfill her duties as a daughter-in-law, nurse the poor and sick of the neighborhood, and adhere to her first spiritual director's rigid advice. After Francis de Sales became her second spiritual director, Jane gradually discovered that it was precisely through the *relationships* with family, her friend Francis, and a wider community, that she would experience the purification that promoted her union with God.[1]

In partnership with Francis de Sales, she eventually founded the Visitation order and developed what has come to be called Salesian spirituality, since Jane's original contributions were only recently noticed and affirmed. Intimate, affective, generative friendship was a central influence in both Francis' and Jane's life and spirituality. Jane lived for almost twenty years after Francis' death, contributing to Salesian tradition her own wisdom regarding prayer, maternal love, gentle leadership and loving community. A significant resource for this wisdom was her strenuous work as a foundress of eighty-nine monasteries, as well as her experience of parenting her grown children and suffering their traumatic deaths.

Jane's and Francis' spirituality emphasizes that one need not be in a monastery to experience the most profound union with God. To "live Jesus" is the heart of the matter, something that everyone can do whether in the French court or on the farm. This death to egotism involved in tender love for each person in community or family is what characterizes their asceticism, in deliberate contrast to the asceticism of

1 Resources for the life and spirituality of Jane include her works collected in *Sainte Jeanne-Francoise Fremyot de Chantal. Sa Vie et ses oeuvres* (Paris: Plon, 1874). Although neither critical nor complete, it is very careful and useful. Her letters are appearing in a critical edition, one volume at a time, done by Soeur Marie-Patricia Burns, *Sainte Jeanne-Francoise de Chantal. Correspondance* (Les Editions de Cerf, 1986). See also Elisabeth Stoppe, *Madame de Chantal: Portrait of a Saint* (London: Faber and Faber, 1962); Wendy Wright, *The Bond of Perfection* (Paulist, 1985); Joseph Power and Wendy Wright, *Francis de Sales and Jane de Chantal: Letters of Spiritual Direction* (Paulist, 1988).

fasting and sleep-deprivation characteristic of all the monastic orders they knew.

Jane's emptiness is profoundly shaped by her experience of widowhood, motherhood, friendship, and relationship to her Visitandine sisters and the men and women she supported as a spiritual director.

Cornelia Connelly (1809–1879)

In barest outline, Cornelia Connelly's life sounds like a script for television's Masterpiece Theater. It encompasses youth as Presbyterian then Episcopalian; adulthood married to Pierce Connelly, a priest who gradually drew Cornelia with him into conversion to Roman Catholicism; motherhood of four children, including a daughter who died at birth and a son who, at age two and a half, burned to death by falling accidentally into boiling juice. She struggled with the consequences of her husband's judgment that he was called to be a Catholic celibate priest and, therefore, to be permanently separated from his wife and young children. While only legal separation was required, Pierce, who wanted to be a Jesuit, urged Cornelia also to enter religious life in England where he moved the family from the United States. Cornelia's own religious vocation slowly emerged in this painful context, and the second half of her life was spent as foundress of a religious society that educated children and prepared teachers to work in anti-Catholic England at the time of the restoration of the Catholic hierarchy. Her religious vocation never mitigated her suffering as a mother who bore the loss of contact with her pre-adolescent children because of Pierce's deception and secrecy.[2]

The turning point of these two acts of her life's drama was her husband's reversal. After his Catholic ordination and Cornelia's novitiate, Pierce discovered he could no longer visit his wife and children in the convent whenever he chose, nor could he control their future as he desired and as he presumed his fatherly role dictated. This led him to denounce Catholicism and sue in English court for restoration of his conjugal rights. Both Cornelia, then a vowed religious in the society she founded, and her children, were regarded in English matrimonial law as

[2] This section is based entirely on Radegunde Flaxman, *A Woman Styled Bold: The Life of Cornelia Connelly 1809–1879* (London: Darton, Longman and Todd, 1991).

the property of her husband, who was the only legal "person" in the marriage. The case dragged on for years, involved press accounts in which Pierce distorted every aspect of the situation (even called Cornelia's convent a "brothel"), and generated such anti-Catholic fervor that Cornelia feared violent attack. The case was dismissed only because Pierce could not pay final court costs. Then, without a word to Cornelia, Pierce took their children out of the country. Cornelia was never publicly vindicated.

While the legal battle was the most public drama, the tragedy continued. In the second act, Cornelia was plunged again into conflict with superior forces. This time it was not the male-dominated, anti-Catholic legal system, but the power of the English and Vatican hierarchies who refused final approbation of her Society, the Sisters of the Holy Child Jesus, for reasons they either withheld from her or used as a "catch 22." Cornelia died without any sign that her life's effort would ever be sealed with the security of Vatican approbation for her Society.

Cornelia's spiritual darkness, like Jane's, was profoundly connected to her experience as a wife, mother, foundress, sister and friend in community. Loving relationships were the primary locus of her emptiness. Like Jane, her personal surrender in contemplative prayer was inseparable from surrender to God's will (their language for emptiness), accomplished in ordinary life in a religious community directed toward the transformation of society, as they understood it.

WHAT COUNTS AS "THE TRADITION" OF SPIRITUAL DEVELOPMENT?

Some scholars claim to notice a radical development in Christian spirituality when they discover the contemporary emergence of three characteristics of spiritual emptiness. First, the goal becomes not the reign of love in one's mystical center, but social transformation. Second, ascetical *kenosis* (self-emptying) is not that associated with contemplative prayer (stilling of mental faculties), but *kenosis* of loving service, "emptying the ego before the other." Third, the traditional order of spiritual growth is reversed so that first there is mysticism, then asceticism or detachment arising from spiritual darkness.

If this investigation had looked to Jane de Chantal or Cornelia Connelly it would have spotted the "radical development" much sooner.

Moreover, noticing this alternative pattern of spiritual emptiness might have prompted a more basic question: Is this a shift from the normative tradition or, rather, another equally significant strand of authentic tradition?

Why would someone interpret the normative tradition as starting with asceticism and only later ending with mysticism? Does this pattern emerge from careful attention to the richness of concrete experience, or from the pattern of organizing the study of spiritual theology in nineteenth century seminary manuals where spirituality was a subdivision of moral theology and concerned first with ascetical theology and then with the mystical way open to those in the "higher" contemplative life of monks and nuns? More careful attention to the contemplative theology of teachers such as Julian of Norwich and John of the Cross would reveal a constant theme: the personal experience of God's love is the only appropriate foundation and motive for asceticism. Or, if one did not begin by attending to theology, one might also notice the pattern of loving relationship as the locus for ascetical self-transcendence by noticing the particular experience of holiness in women such as Jane de Chantal and Cornelia Connelly.

In contrast to the classical pattern which views asceticism as an exercise in detachment that may eventually result in mystical union if God grants this special grace, there is an alternative pattern. It is detachment for the sake of union *now* in love, and both men and women reveal it. Nevertheless, women's experience of it differs from men's. In women such as Cornelia Connelly and Jane de Chantal, asceticism is primarily detachment for *fusion* with relationships since fusion is the most common form of "inordinate attachment" for women. They undergo this detachment for the sake of greater *mutuality, inclusiveness, availability* in relationships, especially in relationship to God. In men, such as Francis de Sales, detachment is primarily from *control* in relationship, since control is the most common form of "inordinate attachment" for men. Men also undergo detachment for the sake of deeper, more inclusive love. In all these cases the detachment or emptiness is precisely for the sake of having the kind of love that God-in-Christ shows in friendship or ministry, a love that supports the other in being her or his authentic self in relationship to God and others. It is not for the sake of release from the suffering that must go with deep relationships. Women like Jane and Cornelia retain a kind of mature "attachment" they do not wish to relinquish because that would be a selfish escape from the pain that inevitably accompanies deep love for friends, spouse, or children.

Another dimension of this issue of the precise *experience* of spiritual darkness is how emptiness *feels* for these women in relation to the way it feels for the men in their lives. For these women, the emptiness of detachment is so often born of the demands of parenting or widowhood that it does not feel at all like the detachment tradition extrapolated from the dominant male tradition that usually describes it as "freedom." Rather, in the life of Cornelia Connelly, for example, it feels like an ache. She said that her Society was founded on a heart breaking with the news that her husband wanted to leave his family and take up a celibate religious life. At other times, Cornelia described her emptiness or detachment as sorrow, as identification with Mary, Mother of Sorrows watching her son's torturous death, as she, Cornelia, bore the pain of her son's slow death from falling into boiling juice.

For Jane de Chantal, detachment felt like nakedness with its accompanying vulnerability and frailty. One description begins with language of "denuding" and "striping," which has associations with voluntary nude posing. Quickly, however, the image changes to one of undefended nakedness and of remaining that way as she experienced very little support to sustain her.[3] Only once did she mention feeling free. All the other feelings were associated with desiring to depend on God alone as her central love, and feeling this love as distant, even absent, and herself as profoundly vulnerable. In contrast, Francis de Sales, to whom she described these feelings, replied that although he, too, found himself "naked," he imagined "...how happy Adam and Eve were when they did not have clothes!" Jane makes no such comment about nakedness making her feel happy.

My aim here is not to make rigid distinctions between men's and women's experience. Rather, it is to invite readers to notice how these examples call attention to the way one's particular experience of being male or female according to society's interpretation of sexuality operates to shape one's experience of spiritual struggle for union with God and others. By noticing concrete experience we can better appreciate and honor the diversity of religious experience.

3 The distinction between nakedness and nude posing comes from Wendy Wright, *The Bond of Perfection.*

Ecclesial Religious Communities *as* Teachers *and* Learners

Ellen M. Leonard, CSJ

The years from 1946 to the present have seen major shifts in both theory and praxis of community and of education.[1] These shifts have had a profound impact on the lives of apostolic women religious. Recently, I taught a course on "The Ecclesial Significance of Religious Life and Emerging Lay Communities." My seminar group included four women religious, two members of male clerical orders, three Anglicans (a woman and man preparing for ordained ministry and a young lay student), and one Baptist woman. My experience with this ecumenical group convinces me that we need to see the issues facing canonical religious in an ecumenical context. All ecclesial communities, with the notable exception of fundamentalist groups, are experiencing diminishment. This diminishment may be interpreted as a purification, a call to something new which is coming to birth. Joan Chittister's image of *grieshog*, the Gaelic process of keeping old fires alive for the sake of building new ones, is an apt one to describe what many persons from a variety of religious communities are experiencing.[2] The experience is one of both pain and promise. What are we learning which might be helpful for the larger church?

I will focus particularly on apostolic congregations of women religious as a case study that may have implications for other types of communities. This essay reflects on some of the ways that women's ecclesial communities have been and continue to be both teachers and learners and how, like all good teachers, they challenge the church in its role as teacher and learner.

1 These are the years during which Margaret Brennan has been a member of an ecclesial religious community of women dedicated to the ministry of education.

2 See Joan Chittister, *The Fire in These Ashes: A Spirituality of Contemporary Religious Life* (Kansas City: Sheed & Ward, 1995).

THE CHURCH AS TEACHER AND LEARNER

There has been a tendency, particularly evident in the past two centuries, to divide the church into the *ecclesia docens* and the *ecclesia discens*, the teaching church and the learning church. The pope and to a lesser extent the bishops are considered the *ecclesia docens* while the rest of the church is the *ecclesia discens*. Such a view obscures the mission of the whole church to be both teacher and learner by overemphasizing the official magisterium and neglecting the role of other teachers/learners in the church. The fact that the official magisterium is composed exclusively of celibate males raises questions concerning their teaching in many areas, but especially in the area of sexuality. As Mary Hines points out, a static understanding of truth allows the magisterium to assume that one small group of the church's members has the truth and the larger and diverse remainder of the community must conform their experience and beliefs to that truth.[3] Other interpretations challenge this inadequate understanding of the way that the Spirit guides the church.

In the last century Newman emphasized the sense of the faithful as "the voice of the infallible Church."[4] At the beginning of this century George Tyrrell strongly objected to the centralization that had taken place within the church which resulted in separating the pope from the rest of the church. Tyrrell insisted:

> at the beginning there was not a teaching church and a learning Church, but a teaching Church and a learning world. Every Christian in virtue of his (*sic*) baptism was a teacher and apostle.[5]

Every Christian must also be a learner.

The emphasis on teaching and learning as functions of the entire church has been affirmed by other theologians, particularly by Yves

[3] Mary E. Hines, "Community for Liberation," in *Freeing Theology: The Essentials of Theology in Feminist Perspective*, ed. Catherine Mowry LaCugna (San Francisco: HarperCollins, 1993), 109.

[4] See Newman's *On Consulting the Faithful in Matters of Doctrine*, ed. John Coulson (London: Geoffrey Chapman, 1961). This work first appeared in the *Rambler*, July 1859.

[5] George Tyrrell, *Medievalism: A Reply to Cardinal Mercier* (London: Longmans, 1908), 62–3. This work has been republished by Burns and Oates with a foreword by Gabriel Daly.

Congar in his important work on *Lay People in the Church*.[6] Vatican II affirmed the active role of the faithful in the articulation of faith, maintaining that the Spirit is present with the whole church.[7] In her study of the women of Vatican II Carmel McEnroy documents how some of the bishops at the council learned to appreciate the role of the laity as "experts in life."[8] Unfortunately this appreciation seems to have waned in the post-conciliar period.

The recognition that the whole church is both teacher and learner does not imply that there are not different functions within the church as well as different ways to teach and learn. As Rahner pointed out,

> Side by side with the official function which is transmitted in a juridical manner, there is and must be the charismatic and the prophetic in the Church which cannot be officially organized right from the start but must, in all patience and humility, be given sufficient room for growth, even though its bearers are sometimes rather "inconvenient."[9]

Among those "inconvenient" charismatic and prophetic persons are founders of various religious orders. One thinks of Mary Ward, Clare, Francis, Ignatius, and countless other creative women and men of vision who continue to teach the whole church through their writings as well as through their followers. Charisms are permanent gifts for the sake of the church and its mission. These gifts have been kept alive, often for centuries, by persons who resonate with the charism of the founder or founders. The charisms have been given institutional form in the orders which continue the work begun by their founders.

6 Yves Congar, *Lay People in the Church*, trans. Donald Attwater (Westminster, MD: Newman Press, 1965), 292–94. See also Leonardo Boff, *Church: Charism and Power*, trans. John W. Diercksmeier (New York: Crossroad, 1985), 138–43.

7 See *Lumen Gentium* 12 and 37; *Dei Verbum* 10; *Apostolicam Actuositatem* 2 and 3; *Gaudem et Spes* 43. For discussion of the teaching authority of the faithful see *The Teaching Authority of the Believers*, *Concilium* 180, eds. Johannes-Baptist Metz and Edward Schillebeeckx (Edinburgh: T. & T. Clark, 1985).

8 Carmel McEnroy, *Guests in their Own House: The Women of Vatican II* (New York: Crossroad, 1996).

9 Rahner, "Freedom in the Church," *Theological Investigations* 2 (Baltimore: Helicon, 1963), 107.

Religious life rightly belongs to the charismatic and the prophetic, in spite of efforts to control the gifts of the Spirit by emphasizing the juridical. How have religious congregations exercised their charismatic and prophetic role as teacher/learner within the church and society—and how might they do so today?

ECCLESIAL RELIGIOUS COMMUNITIES AS TEACHER/LEARNER

The renewal of religious life initiated by Vatican II was part of the renewal of the whole church. Apostolic religious life shifted from a stance of renunciation of the world to a prophetic witness within the world in order to work for its transformation. As ecclesial bodies, religious congregations moved from a triumphal image of people set apart and immune to the world, to a people who are in *solidarity* with the whole struggling people of God, and with all humankind in its fragility and pain. Today we include not only humankind but all of creation. In her account of the women of Vatican II McEnroy includes a chapter on "The Nun in the World" in which she describes the process of change. It was "more than a tailor's one-night job," but the change in dress dramatically symbolized the shift in self-understanding that took place as apostolic religious discovered by trial and error how to live their religious lives in the modern world.[10]

In response to the directives of Vatican II religious congregations returned to their roots—to their particular founding visions. They discovered the story of their beginnings. Often a few people had been willing to dream dreams, to take the gospel message seriously, and to defy opposition from many quarters, even from church authorities, in order to respond to what they discerned as God's call to meet a particular need.

In the years preceding Vatican II large numbers of generous young women entered religious life. Perfection was the goal and detailed rules guided the novice along this path. One learned how to be a teacher and then taught others. The change from "state of perfection" to a "pilgrim people" on journey with all the baptized has had important implications for ecclesial communities of teaching and learning. How one journeys into an unknown future is very different now than it was when the

[10] *Guests in Their Own House*, 161–97.

novice was presented with a detailed map. The life itself, as well as the concepts of teaching and learning, has changed dramatically. But change is not new for religious congregations. History presents many different forms that religious life has taken.

RECALLING THE PAST

Religious communities have always been places of teaching and learning. From the sixth to the sixteenth centuries women found opportunities for education in convents and monasteries. One result of the Protestant Reformation was the loss of female centers of culture and learning with the suppression of monasteries. Janice Raymond points out that at the time of the Reformation some female convents were turned into male colleges of higher learning; for example, the nunnery of St. Radegund's in Cambridge became Jesus College.[11] Although the opportunities provided by monasteries were limited to the privileged few, the presence of spheres of female autonomy and learning was significant for women who were excluded from the universities.

In response to the Reformation the Council of Trent initiated the reform of religious life by legislating a return to monastic life with solemn vows and strict enclosure as the only "approved" form of religious life for women. But while the Council Fathers insisted on the cloister, women were developing new ways to exercise discipleship in response to the pressing needs of church and society. In small groups they were banding together in a new form of apostolic religious life. One of the needs to which they responded was the education of girls. The nuns in their monastery classrooms, double-locked so as not to violate cloister, showed that the education of girls was an important way to influence families, for girls would become wives and mothers. While the daughters of wealthy men could be educated in a monastery, there were no schools for poor children. The Catholic Church did not approve of co-education, but recognized the value of educating future mothers who would pass on their faith to their children. The new communities could provide dedicated teachers to uphold the Catholic faith.

The growth of congregations of apostolic religious women devoted to teaching which took place from the sixteenth through to the twenti-

11 Janice G. Raymond, *A Passion for Friends: Toward a Philosophy of Female Affection* (Boston: Beacon Press, 1986), 105.

eth centuries is remarkable.[12] Church authorities were initially suspicious of the women who went out to teach and, unlike "real" religious, did not observe rules of enclosure. Teaching was considered a male profession and an unsuitable task for a woman, one which would be a threat to her chastity. Gradually, dedicated Catholic women broke down resistance to female teachers by providing a needed service in the community. Besides, schoolmistresses were less costly than schoolmasters, and in fact the new congregations did not charge their students. Schooling was free, the sisters supporting themselves and their schools by their handwork. Even stronger objection was made to women catechizing, a task which belonged to the clergy, but the needs were so great that eventually the schoolmistresses won a place within both church and society. The teaching congregations flourished and became indispensable. Apostolic congregations provided the education necessary to meet the challenges of the day and to support Counter-Reformation Catholicism.

In keeping with the gender perspectives of the time, education for girls focused on preparation for motherhood and home management. To seek learning for its own sake was considered unfeminine and even dangerous. Religious instruction was emphasized but reading was also taught since it allowed access to the word of God. Fewer girls learned to write, a task which required tools and facilities. Some moralists saw writing as a dangerous skill for women, one which would enable them to carry on underhanded liaisons. Handwork was considered useful for girls of all social classes. Such work provided a way that women could avoid "the evil of idleness," and working class and poor girls learned manual skills which enabled them to support themselves. Girls usually stayed in school for only two or three years, until they received their first communion at the age of ten or twelve.

Not only did the religious congregations educate girls throughout Europe, but they also volunteered to go to the mission fields opening up in the Americas and other parts of the world. The teaching congregations became instruments in the rechristianization of society in Europe and pioneers in the New World, where they exercised a powerful influence on Catholics and Protestants through their work in schools.[13]

12 For an account of the development of teaching congregations in France see Elizabeth Rapley, *The Dévotes: Women and Church in Seventeenth-Century France* (Montreal: McGill-Queen's University Press, 1990).

13 See Mary Ewens, "Leadership of Nuns in Immigrant Catholicism," *Women*

Although their lives were often difficult, they enjoyed opportunities for education open to few other women of their time. Apostolic congregations of women continued this commitment to education, especially for girls, providing education from kindergarten to college. They not only educated future mothers who would teach their children, but also the next generation of teachers.

Both the content and the methods used in teaching were influenced by the particular charism of the religious congregation to which the teachers belonged. Lorna Bowman offers an interpretive framework within which to clarify the relationship between the religious and the educative dimensions of the educational philosophy of Catholic religious congregations committed to education. Her work has focused on the Society of the Holy Child Jesus founded by Cornelia Peacock Connelly, but it has implications for all congregations involved in education.[14] Teachers and students alike imbibed the charism of the founder, which was then reflected in their lives.

PRESENT CHALLENGES

The kind of teacher/learner required in our day is very different from the absolutist authoritarian models of the past.[15] Knowledge itself is so vast that individuals need access to networks and to communities of learning. Teaching and learning are communal activities. Persons who have chosen to throw in their lot with a community have had rich experiences of learning how to learn together.[16] Marcel Dumestre maintains

and Religion in America, Volume 1: *The Nineteenth Century*, eds. Rosemary Radford Ruether and Rosemary Skinner Keller (San Francisco: Harper & Row, 1981), 101–107; Asuncion Lavrin, "Women and Religion in Spanish America," *Women and Religion in America*, Volume 2: *The Colonial and Revolutionary Periods*, eds. Ruether and Keller (San Francisco: Harper & Row, 1983), 42–50; Christine Allen, "Women in Colonial French America," Ibid., pp. 79–86.

14 "The History of Women, Religion and Education: A Methodological Approach," *Toronto Journal of Theology* 11 (Fall 1995): 201–16.

15 Some of the changes are explored by Frances A. Maher and Mary Kay Thompson Tetreault in *The Feminist Classroom* (New York: HarperCollins, 1994).

16 *Women Religious and the Intellectual Life: The North American Achievement*, ed. Bridget Puzon (San Francisco, London: International Scholars

that small intentional learning communities which foster trust and commitment offer a new paradigm for effective adult Christian religious education.[17]

Religious communities have been and continue to be communities of learning. The setting may not be a school, although schools are important centers for learning. Other settings include retreat houses and spirituality centers throughout Canada and the United States, where new and innovative forms of teaching and learning are occurring. Some of these communities have a strong ecological commitment and are experimenting with new ways of using resources.

Religious life has always offered a viable alternative, especially for women. Today more than ever our church and society need alternative communities where persons learn to live in communion with all—communities of hope, compassion, and friendship. There is a deep longing for connection, meaning, and belonging. Often the longing takes the form of what Schillebeeckx calls a contrast experience, a sense of the lack of connection, meaning, and belonging. The experience of disconnectedness can enable us to recognize our call to communion.

Letty Russell's "church in the round" offers an image of church with which many congregations of women would feel at home.[18] The "church in the round" is a place of hospitality for all, especially for those on the margins of both church and society. It includes hospitality to one's self in what Russell calls "sister choice."[19] This choice for oneself as a woman and for other women is one of the learnings that women religious have discovered during the past thirty years, one which they try to share with the larger church. Women in religious congregations have enjoyed a certain status within the church that other women have not had. As we learn to recognize that women and men are equal disciples

---

Publication, 1996) studies the contribution to the intellectual life and future potential of women's religious communities. The title is misleading since data is drawn only from religious congregations of women in higher education in the United States. The Canadian experience was not considered.

17 Marcel Dumestre, "Postfundamentalism and the Christian Intentional Learning Community," *Religious Education* 90 (Spring 1995): 190–206.

18 Letty M. Russell, *Church in the Round: Feminist Interpretation of the Church* (Louisville: Westminster/JohnKnox Press, 1993).

19 Ibid., 183–87.

we join in solidarity with all women, especially those who are oppressed. Our sense of sisterhood is being expanded as we learn to overcome our own sexism, racism and classism.

The traditional vows of poverty, obedience, and chastity, when freed from rigid juridical interpretations often used as a form of control, may provide a source of wisdom. One of the lessons that religious congregations are learning as they struggle to live evangelical poverty is that most of us in North America can do with less. Joan Chittister suggests a spirituality of diminishment is needed today.[20] Those who have tried to live simply in the midst of affluence may have some wisdom to offer at this time of economic disparity.

Religious communities are struggling to find ways of living collaboratively. Many have discovered non-hierarchical ways of exercising leadership and have devised processes which free the creativity of the individuals in the community. All members share in the task of discerning their future together and of living in obedience to the Spirit. Such wisdom can enrich the larger church.

Perhaps in our day one of the most important learnings is in the area of personal relationships. Women religious are learning how to live in non-exploitative intimate relationships. This is a lifetime challenge. The traditional vows of poverty, obedience, and chastity may be seen as charisms—gifts for the larger church—which open the community to new ways of sharing resources, new ways of collaborating for mission, and new ways of entering into personal relationships.

A new language is developing which is changing how we learn, and religious communities need to learn it, a fact John Paul acknowledged in his recent Apostolic Exhortation, *Vita Consecrata*:

> Consecrated persons, especially those who have the institutional charism of working in this field, have a duty to learn the language of the media in order to speak effectively of Christ to our contemporaries, interpreting their "joys and hopes, their griefs and anxieties," and thus contributing to the building up of a society in which all people sense that they are brothers and sisters making their way to God.[21]

20 *The Fire in These Ashes*, 71–74.

21 John Paul II, *Vita Consecrata*, 99 in *Origins* 25 (April 4, 1996): 681–719.

Cyberspace is rapidly changing the way we interact with one another. New possibilities for networking are providing links between persons and communities on a global scale. The popular "Sister L" on the internet enables persons from all continents who are interested in religious life to exchange ideas and offer support to one another. It includes women and men from many different congregations, academics from various disciplines, and other interested persons. When I met a Jewish nurse from a kibbutz in Jerusalem wearing a "Sister L" shirt, I recognized that a new form of community is emerging, one which is ecumenical and interfaith. Through technology we are experiencing our interconnectedness in new ways.

The distinction between teacher and learner vanishes as the traditional teacher at the front of the classroom disappears. We are living on new boundaries which are challenging our traditional ways of thinking. As we bravely enter into this electronic age we need companions, persons who are committed to one another and to a vision of the world in which all life is valued.

The present may be a new charismatic moment in the history of religious congregations. We are invited to respond to the gospel call to discipleship not just as individuals but as communities, and to do this in ways that address the larger community—church and world. The future of ecclesial religious communities is uncertain. The challenge, as Chittister articulates it, is "to live this time now, our time, well so that a future model can rise from these ashes with confidence and with courage."[22]

Perhaps the greatest lesson that ecclesial religious communities can provide is a commitment to continue the search for God. The search is never over. It cannot be undertaken alone. Ecclesial religious communities provide a space in which that search for God can continue. Religious life is a particular actualization of the call to discipleship, a call rooted in our baptism. It provides an alternative way of living the Christian life, or in the words of Jon Sobrino, "of becoming Christian."[23] Its contribution to the mission of the church flows from its fidelity to its own charism. By keeping alive the mystical and the prophetic dimension of

22 *The Fire in These Ashes*, viii.

23 Jon Sobrino, *The True Church and the Poor* (New York: Orbis, 1984), 302–37.

the Christian life, ecclesial religious communities have valuable resources to share. In ever new ways they continue the mission of the church to be always both teacher and learner.

Issues *in the* Understanding *and* Exercise *of* Authority *in* Ecclesial Religious Institutes

Sharon Holland, IHM

Moses is not an everyday saint. Still, religious superiors (by whatever name) may well find inspiration, strength and even comic relief in the monumental figure of Moses trying to lead Israel forward. Apparently chosen by God, the reluctant Moses finds the people rallying around him at one moment and stumbling and grumbling against him the next. He talks to God face to face in plain words, reminding God: "these are *your* people." When they grumble about hunger, Moses intercedes for food; when they grumble from thirst, he negotiates water. When they turn to apostasy, he intervenes against God's wrath. In the dramatic scene of battle against the Amalekites, Moses' close collaborators Aaron and Hur must keep his out-stretched arms propped up as he implores God's favor for Israel. In the end, an aged Moses sees that the goal will be attained, but not by him; his young successor Joshua will bring the people into the promised land.

The story of Moses portrays humanity and divinity, fidelity and infidelity; it speaks of his daily encounters with God and with the community. There are moments which touch the depths of desperation and the peaks of glory.

Every religious leader must know such moments. However, it is only in the present century that general superiors of sisters' and brothers' congregations have been recognized as even possessing the authority of their office. On the brink of a new century, following extraordinary changes, it is crucial that they not inadvertently relinquish the ground which has been gained.

The considerations which follow are divided into two parts. In the first is an analysis of some of the canonical developments since 1900. Following this are some reflections on the actual situation as this century draws to a close.

Canonical Developments in the Twentieth Century

During the first year of this century, Pope Leo XIII promulgated a document of enormous significance for apostolic women religious. The apostolic constitution, *Conditae a Christo* (8 December 1900), provided for the first time real authority to superiors general of these congregations. Although this reality is now largely taken for granted, many contemporary religious know the history of divisions and collisions occasioned by moves or attempted moves into new dioceses, or by attempted decisions by superiors which were not then fully within their authority.

For centuries, religious were, by definition, members of the orders pronouncing solemn vows. The great number of religious families which grew up in response to needs of the Church and world, most dramatically after the French Revolution, pronounced simple vows, were free of strict cloister in order to pursue their apostolic purposes, and were variously called sodalities, pious unions or associations. Some were even called congregations.

The fact that *Conditae a Christo* recognized as religious these congregations with simple vows dedicated to works of mercy, is perhaps of less significance. These men and women (both brothers and sisters are the subject of the document) were widely seen and recognized as religious in view of their vows, common life, habit and good works. What was crucially lacking was that general internal power of governance which today's Code calls "rightful autonomy" (c. 586). As a result of Leo XIII's action, the so-called sodalities would henceforward be numbered among religious with public vows and, distinct from the orders, would be known as religious congregations or simply congregations. Some of these having been in some way recognized by the Holy See would be known as pontifical. Experience had shown that when these spread to various dioceses, certain difficulties arose. Thus it was recognized that the time had come to place certain limits on the authority of the bishops in order to allow congregations with simple vows to have a unified government under their own general superior. The jurisdiction of the bishop according to the canons and the apostolic constitution would remain, but these congregations were not diocesan, and while subject to the bishop's authority in his diocese, this was not to prejudice the administration and government of the general superior.

The text of *Conditae a Christo* cannot be read today without thinking of *Mutuae relationes* (1978). In recognizing a real authority to be exercised by general superiors, the apostolic constitution realized a new need to coordinate these two legitimate authorities. General superiors must not infringe upon the rights and power of bishops; bishops are not to take over that authority which pertains to the superior.[1]

The two sections which follow the above statement treat pontifical congregations and those not yet recognized by the Holy See. In pontifical congregations, the general superior now had the right to accept candidates and to admit them to the habit and profession of vows. Although the tridentine norm for an episcopal examination before the reception of the habit and profession was retained in the case of women, the right of admission now belonged to the superior. Likewise, it was the superior's role to organize individual communities and to dismiss candidates and professed according to the norms of law. Dispensation was reserved to the Holy See. Chapters and councils had their own proper role regarding the whole congregation and the individual houses—retaining the right of the bishop to preside at the equivalent of the elective Chapter in the case of women's institutes.[2]

By way of contrast, in diocesan groups, it remained the diocesan bishop's right to admit to reception of the habit, and to profession of vows. He had the authority to dismiss or dispense, though not without the knowledge of the moderator. In these congregations, members could not go from one diocese to another without the permission of both bishops. Finally, although these sisters had the right to elect their superior, the bishop retained the right not only to preside, but to confirm or rescind the election.

To ensure its provisions, the document extended protection to the constitutions of pontifically recognized groups. Once they were approved by the Holy See, the bishop had no right to change the constitutions. Likewise, the governing authority over the whole institute or its

1 Leo XIII, apostolic constitution *Conditae a Christo* (8 December 1900) in *Codicis Iuris Canonici Fontes*, Petri Card. Gasparri, ed. (Rome: Typis Polyglottis Vaticanis, 1925) Vol.III, 562: "Qua igitur ratione summis societatum harum Praesidibus in Episcoporum iura et potestatem nefas est invadere; eadem Episcopi prohibentur ne quid sibi de Praesidum ipsorum auctoritate arrogent." The 1978 document commonly known as *Mutuae relationes*, was prepared jointly by the Sacred Congregations for Bishops and for Religious and Secular Institutes in order to provide doctrinal and pastoral reflections to guide the relationships between bishops and religious.

2 Ibid.

individual communities which, according to the constitutions belonged to the superiors, could not be altered by the bishop.[3]

Institutes not yet pontifically recognized enjoyed a certain protection in that their nature and law could not simply be changed when they moved to a new diocese. Such changes required the consent of the bishop of each place where they had houses. This, however, was surely a part of the difficulty which gave rise to *Conditae a Christo*.

Also in the realm of governance, it was recognized that the administration of temporal goods belonged to superiors with their councils.

The Bishop's usual role of vigilance with regard to the apostolate is presented, along with the right of visitation. The latter is broader in the case of brothers and sisters than in instances of congregations of priests in simple vows. With regard to internal governance of institutes, however, those recognized by the Holy See were enabled, through the concessions of *Conditae a Christo*, to enjoy a much more stable unity through the office of a general superior with real authority throughout the institute.

Today's emphasis on founding charisms and fidelity to a founding spirit, in the midst of extensive inculturation, serves to highlight the importance of this unity of government throughout an institute. Pope Leo XIII himself, in a 1901 letter to superiors general, urged a faithful following of founding inspirations. Here was a new possibility for universality, rather than diocesan disunification.

The Code of 1917 incorporated the provisions of *Conditae a Christo*, recognizing for such superiors "dominative power" (c. 501) of a private nature. Significant studies have been published on the evolution from that 1917 canon to the 1983 Code's treatment in can. 596, together with its implication for lay sharing in the *potestas regiminis*. This important point is not the direct focus here. Rather, it is to explore further the potential of the office of religious superiors—especially that of general superior—building upon the foundation of *Conditae a Christo*, and developing its potential for: 1) unifying a congregation internally according to its particular charism no matter how extensive its geographic spread, and 2) fostering ecclesial communion.

Pope Leo XIII, in order to provide a unified government, provided

3 Ibid., 564: "Item regimen, quod penes moderatores est sive sodalitatis universae sive familiarum singularum ad constitutionum normam, Episcopis mutare temperare ne liceat."

the foundation through recognition of certain pious societies of simple vows *sub regiminis Moderatoris generalis*.

The 1983 Code for the Latin Church recognizes the authority of the "supreme moderator" for all provinces, houses and members (c. 622). It is an earlier canon, however, which reiterates the evangelical and ecclesial dimensions of this role. Canon 618 begins with a statement on the source of authority in religious institutes and continues with a development of its various roles. We read: "Superiors are to exercise their power, received from God through the ministry of the Church, in a spirit of service."

In the case of *Conditae a Christo*, it is clear enough that the Pope, through an apostolic constitution, recognized for congregations a new authority to be secured in proper law. Furthermore, he explicitly limited the authority previously held by the diocesan bishop with regard to the same institutes and persons. There was, as it were, a shift in the balance of power within the Church for the good of the institute's internal communion and, to use today's terminology, it was to be exercised in full ecclesial communion. The granting of this power appears essentially one of juridic recognition on the part of the Pontiff, but in his above-quoted letter, he also extolled fidelity to diverse founding charisms. Commentators on the 1917 Code suggested that the "dominative" power of these superiors flowed from religious profession of vows.

It is of significance, then, that the first circulated draft of what would become canon 618 in the 1983 Code explained the source of a religious superior's authority simply as coming from God.[4] Following the broad consultation on the 1977 draft, the commission meetings reflect an initiative to clarify the nature or source of the power under discussion.

In April 1979, one consultor insisted that the power of superiors was in some way ecclesiastical, being derived from ecclesiastical power; it was not simple friendly or private authority, or the outdated notion of dominative power. Not all agreed with trying to incorporate this notion however, lest the end result might seem to equate superiors' power with hierarchical power of governance.[5]

4 Pontificia Commissio Codici Iuris Canonici Recognoscendo, *Schema canonum: De Institutis Vitae Consecratae* (Vatican City: Typis Polyglottis, 1977) norm 26: "...a Deo conceditur."

5 *Communicationes* 11 (1979), 306.

One can imagine that consultors, during their sessions or in private study, had returned to *Lumen gentium*'s clarification that institutes of consecrated life do not constitute an intermediate stage within the hierarchical structure of the Church (LG 43). Rather, being of the charismatic dimension of the Church, the gifts of founders and foundresses are confirmed and authenticated by the hierarchy (LG 12).

Several months later, the commission returned to the canon. Again there was the express desire to acknowledge that religious superiors' authority is received from the Church. The consultors struggled to coordinate the concept that all authority comes from hierarchical authority, with the understanding that in some way authority also comes from the charism of the institute. The discussion reflected that recognition as a form of public authority in the Church, and implied some sanction of it on the part of the hierarchy. A proposal was offered: authority is received from God through the Church. The text finally resulting from this meeting recognized a power received from God through the ministry of the Church.[6] The concept of exercising this power in a spirit of service had been a constant in the text, but now was deliberately moved to a more prominent place at the beginning of the canon.

The contemporary importance of these reflections is not primarily, or at least not exclusively, one of *potestas regiminis* and jurisdiction, although it is of considerable importance that that discussion continue. The two-fold but unified source of authority is the same as the origin of the institute itself. The founding charism comes from the Spirit to an individual for the Church, and its authenticity is confirmed by the hierarchy of the Church. A superior exercises an authority formed by the founding spirit and must remain faithful to it. He or she must foster unity within the spiritual heritage of the institute, while guiding the fulfillment of its mission within the broader Church. Thus, there is the role of fostering communion within the institute and communion of the institute with the local and universal Church.

Commentators on these roles frequently speak of the spiritual dimension of this authority. It has its origin in God because the institute itself arises from a gift of the Holy Spirit; the authority exercised is of the spiritual order, as it guides the institute and its members toward the

6 *Communicationes* 12 (1980), 145–146: "...potestatem a Deo per ministerium Ecclesiae receptam."

fulfillment of the purpose for which the gift was given.[7] J. Beyer calls this emphasis on the spiritual aspect of government the most significant innovation in this section of the revised canons.[8]

Some years before the Code, however, *Mutuae relationes*, collaboratively prepared by the curial dicasteries for religious and for bishops, had made a significant contribution. In article 13 on the "Service Characteristic of Religious Authority," the authors found it useful to employ the language of pastoral ministry—teaching, sanctifying and governing—recognizing that all of the People of God have in common the prophetic, priestly and royal conditions.

Under the rubric of the "office of teaching," religious superiors are recognized as having "the competency and authority of spiritual directors in relation to the evangelical purpose of their institute." This suggests on the part of superiors a profound knowledge of the spirit and founding intention of the founder or foundress, and an intimate personal identity with the entire spiritual heritage of the institute. This, coupled with sound knowledge of Church teaching and contemporary world realities, will enable a superior to guide and direct the institute in fidelity to its charism toward the fulfillment of its particular mission.

Obviously closely related is the "office of sanctifying." At the heart of this is fostering growth in charity and communal and individual fidelity in the practice of the evangelical counsels, always in accordance with the end of the institute. In a wide variety of ways superiors can foster this growth, inspiring, encouraging, providing opportunities and materials, and giving example through a life of personal and liturgical prayer and rootedness in faith, hope and love. The superior's own personal life and concern for each member, after the example of Jesus, contributes much to the fulfillment of this aspect of his or her role.

Finally, there is the "office of governing" which, in a spirit of service, is dedicated primarily to providing for the order and organization needed for promoting the life and mission of the institute. In caring for and developing the mission of the institute, the service of governing includes seeing that the institute's particular mission is "efficiently inserted into ecclesial activity under the leadership of the bishops." In the document specifically dedicated to promoting better mutual rela-

7 Velasio DePaolis, *La Vita Consacrata nella Chiesa* (Bologna: Casa Edizioni Dehoniane, 1992), 187.

8 Jean Beyer, *Le Droit de la Vie Consacrée* (Paris: Editions Tardy, 1988), II, 23.

tions between religious and bishops, the rightful autonomy of institutes—now codified in canon 586—is clearly recognized.

A useful summary of the purpose of religious government is provided in the 1983 document *Essential Elements*, which draws on various ecclesial documents. It notes that the principles of personal religious authority and shared responsibility must be rightly understood and implemented so that the purpose of religious government may be fulfilled by "the building of a united community in Christ in which God is sought and loved before all things, and the mission of Christ is generously accomplished."[9]

Belgian canonist, J. Beyer, who stresses the spiritual dimension of religious authority, notes an apostolic priority in the canonical definition of a general superior (c. 622). Global authority and responsibility assure unity of life and mobility of action in institutes dedicated to apostolic works.[10] This is precisely what was lacking prior to 1900, resulting often in fragmentation or simply unwanted divisions within institutes once they spread beyond diocesan borders. The provisions of *Conditae a Christo* opened the way for the flourishing of large, centralized, often international institutes of apostolic sisters and brothers.

CONTEMPORARY REFLECTIONS

In every era, new challenges to the effective use of this authority appear. The study of religious life in the United States, carried out over a three year period by psychologists Fr. David Nygren and Sr. Miriam Ukeritis, concludes that authority "is perhaps the most pressing question to resolve."[11] Sr. Janet Ruffing, RSM, addressing the 1993 LCWR Assembly reiterated the difficulty which results from differing interpretations of authority within a congregation, "impeding the ability of leaders to lead and making it difficult to hold members accountable."[12]

9 Sacred Congregation for Religious and Secular Institutes, *Essential Elements in the Church's Teaching on Religious Life* (Boston: Daughters of St. Paul, 1983), n. 52.

10 Beyer, 31.

11 "Future of Religious Orders in the United States," *Origins* 22 (24 September 1992), 271.

12 Janet Ruffing, "Enkindling the Embers: The Challenge of Research on Religious Life," LCWR National Assembly 1993 (Printed Text), 11.

One of the clearest recent presentations of this dilemma appears in the 1991 publication of sociologist Patricia Wittberg, S.C., *Creating a Future for Religious Life*. In the context of discussing the "associational model" of community, Wittberg notes some of that system's weaknesses. Among others, there is a lack of mission mobility and availability. The capacity for witness is also limited; this model, notes Wittberg, does not authorize its leadership "to challenge members to a counter-cultural stance." An associational congregation cannot rely strongly on the members' commitment and co-operation. That model challenges its leaders' right to exercise spiritual leadership. Beyond this, it also challenges the right to oversee finances and ministry placement.[13]

Obviously, these factors strike at the heart of religious governance: the promotion and facilitation of the institute's life and mission. Speaking from the experience of still another discipline, Theresa Monroe, RSCJ, of the JFK School of Government at Harvard, cites the facilitation of task accomplishment as the first purpose of authority structures. These have a primary function, Monroe observes, of managing organizational boundaries, containing chaos and balancing dis-equilibrium, providing a "holding environment" for risk-taking.[14] In contrast, the author points out that some contemporary processes sap energy and creativity. Many communities have, in fact, "lost the ability to make definitive corporate decisions and take collective action."[15]

Similarly, the Nygren-Ukeritis report recognizes that the absence of a corporate commitment which would enable a response to unmet needs "in the light of Gospel imperatives" stands in contrast "to the collective vision and action, rooted in God which marked the birth of most...congregations."[16] While this obviously requires the willing responsiveness of all members, the study's profile of effective leaders includes their clear sense of the impact a congregation could have, and their understanding of how to position the congregation for response to human needs.[17]

13 Patricia Wittberg, *Creating a Future for Religious Life* (New York/Mahwah, N.J.: Paulist Press, 1991), 74–79.

14 Theresa M. Monroe, "Reclaiming Competence," *Review for Religious*, 51 (1992), 448–49.

15 Ibid., 435.

16 "Future of Religious Orders," 270.

17 Ibid., 271.

Noting that dramatic change is essential to religious life in the U.S., without which it will continue to decline and those "most in need of their help will not be cared for...", the report highlights the importance of leadership in effecting transformation. The authors stress "the urgency of selecting and training leaders who not only can manage the complexity of religious life...but who also can focus the attention of their communities on a vision that will unite individual efforts inspired by the mission of their founder or foundress."[18]

In the broader context, Ruffing points out the important contribution religious could make through learning to exercise authority within a community of adult disciples—exercising power and authority as Jesus would.

A century ago, the general superiors of sisters and brothers with simple vows simply did not have the authority suggested by their title. Progressively through the apostolic constitution *Conditae a Christo* (1900), the 1917 Code of Canon Law and the 1983 Code, their authority to govern their religious institutes with rightful autonomy has been consolidated and enhanced. In the second half of the twentieth century the establishment of Conferences of Major Superiors and the stress on mixed commissions of bishops and religious superiors has furthered the possibilities for collaboration and communion in the Church.

While there remain serious questions to be resolved, the firm foundation of a century is there to be built upon, a foundation which can truly serve to unify members in charity across expanding cultural diversity, and extend and fulfill the institute's mission in the Church and the world.

Obviously, this requires mature collaboration on the part of all, but there also must be courageous leadership in the trek across the contemporary wilderness. Those chosen for the task must resist the temptation to abdicate their authority, whether in the name of totally shared responsibility or in the face of modern-day "stumbling and grumbling." Keeping the vision before members' eyes and the promise ringing in their ears, religious leaders will urge their communities to a renewed sense of mission. Like Moses, they will speak familiarly with God. They will lead their Congregations into the third millennium, accepting again the mandate of the risen Jesus, "Go, teach all nations," and relying confidently on his promise, "Behold I am with you always."

18 Ibid., 266–67.

The Desire for God *and the* Transformative Power *of* Contemplation

Constance FitzGerald, OCD

Introduction

All around us today we see a passion to touch the roots of contemplation or mysticism in our history as a people, to hear a muted desire that has existed often only as a subterranean force and to bring it above ground into the public forum in order to understand its power for transformation in our post-modern world.[1] Even if it is unrecognized and therefore uninterpreted, the desire for God is apparent everywhere in so many different forms. If we are able to reclaim this muted desire that runs through our history and make available centuries of contemplative tradition, Carmelite in this case, the dominant paradigms of this tradition may offer some guidance to our nation, North America, the Western world, called as it is by history and so-called "development" to a contemplative time, challenged to mature beyond being first, beyond being the Center of the world. I often feel that only if we are prepared for transformation by contemplation[2] and thereby given a new kind of consciousness and imagination will humanity and the earth, with its various eco-systems, survive.

What the Carmelite contemplative tradition reveals is women and men searching for God, desiring God together. Contemplation cannot be understood except within the context of desire, that is, divine desire coming to meet human desire and igniting in human hearts an

1 The ideas in this essay were first presented in the keynote address at *Carmel 200: Contemplation and the Rediscovery of the American Soul*, held at Loyola College, Baltimore, August 12–18, 1990, to celebrate the bicentennial of Baltimore Carmel and Carmel in the U.S. This symposium was seen as an enterprise of imagination. It sought to retrieve the personal, relational, and social past of the desire for God, particularly in the 800 year old Carmelite tradition, in order to bring the too often marginalized mystical tradition into conversation with our present reality as North Americans and write a new "text" indicative of the future horizons of contemplation.

2 Contemplation has various meanings. In the Carmelite tradition it is understood as a love experience which is also a deep knowing. Hence, John of the

unquenchable desire. Those we today call mystics have always upheld the primacy of desire. Their writings are suffused with desire. They know we are propelled by our insatiable desire and keep trying to tell us religion is a message of desire and a hope of its fulfillment. Teresa of Avila and John of the Cross in their classic works tell the story of human desire and delineate how it grows, is educated and purified, and finally, transformed within the life journey itself. Together, woman and man bear witness to the agony and ecstasy of burning desire. This is why after four centuries their writings still educate us to contemplation.

If the contemplative voice is ever really heard again in theology and the official Church, as it was before the importance given to religious experience gradually separated the patristic theology as found in the monasteries from the theology of the schools, then the primacy of rationality will, according to Brazilian theologian, Maria Clara Bingemer, have to yield a place to "the impulse [passion] of desire that dwells at the deepest level of existence..." Bingemer accents the place desire must have:

> Theology, which seeks to be reflection and talk about God... cannot but be moved and permeated throughout its whole extent by the flame of desire. At a particular point in its theological articulation, reason, science and systematic rigor have their role..., but they can never suffocate the greater desire, never tame the divine pathos, which, from all eternity, has broken silence and become a loving and calling word, kindling in its turn in the hearts of humankind an irresistible and insatiable desire... Born of desire, theology exists as theology only if it is upheld and supported by desire, in the direction of the desire that is its goal and its horizon.

Bingemer maintains, furthermore, that the future of women doing theology is inseparably linked with desire:

> A woman finds it unthinkable to divide her own being into watertight compartments and treat theological work as a purely rational activity. Moved by desire, a totalizing force, she does theology

Cross speaks of a secret, loving knowledge or the loving inflow of God or dark contemplation or dark night. When John says the dark night is an inflow of God, this inflow is very precisely in terms of secret Wisdom, who is Jesus Christ.

> with her body, her heart and hands, as much as with her head, and the ripe fruit she begins to make available is the result of slow and patient pondering of experiences lived deeply and intensely [in dialogue with the tradition]... When we talk about desire we are talking about human beings at their deepest level, in their deepest and ultimate truth, in their vital force, and therefore in their most authentic and legitimate aspirations. We are talking about what makes our bodies quiver and tremble with pleasure, about our noble and threatened vulnerability, our greatness, which depends on our fragility...[3]

Bingemer concludes by suggesting that the challenge to women doing theology today is to restore the primacy of desire within theological discourse,[4] and her challenge suggests a threefold intersection of contemporary feminist theology, the effort to reclaim and reinterpret the mystics, and the experience of the desire for God that pervades this country often unbidden and unseen.

3 Maria Clara Bingemer, "Women in the Future of the Theology of Liberation," in Marc H. Ellis and Otto Maduro ed., *The Future of Liberation Theology, Essays in Honor of Gustavo Gutierrez*, (Maryknoll, New York: Orbis Books, 1989), 478–79.

4 Margaret Brennan is an example of a woman theologian who has understood this challenge in her life and scholarship. In 1969, she entered the world of contemplative nuns by attending the Seminar for Contemplative Sisters held at Woodstock, Maryland, the former theologate of the Jesuits of the Maryland Province. At that time, as leader of her congregation, she was beginning the House of Prayer movement and came with deep appreciation for the life of prayer and the mystical tradition. She supported the fledgling efforts of the Association of Contemplative Sisters founded at Woodstock and continued as President of the Leadership Conference of Women Religious to assist Contemplative Communities in both finding their voice in the Church and renewing their way of life. This was not a comfortable position to assume since, at that time, the Congregation for Religious was discouraging apostolic religious from assisting contemplatives even though the latter had no corporate channel of communication with the Holy See. Outstanding among and typical of her many interventions on their behalf was the address given to the Canon Law Society of America on October 7, 1975. It was fitting that she should close the *Carmel 200* symposium with a presentation entitled "Contemplation Finding Its Prophetic Voice in the Cultural Context of North America." When the history of these years is fully told, Margaret Brennan's contribution to the

WHERE HAVE YOU HIDDEN MY BELOVED?

In 1984 an article appeared interpreting John of the Cross' teaching on desire and dark night in relation to contemporary experience.[5] This present essay reflects further on the societal aspect of the Dark Night because I sense that our consciousness has changed and as persons and as a nation we are in a different place than we were in the early eighties.

For one thing, *signs of death* seem even more pronounced than they were then, while *signs of new life*, new vision, and a new voice, unavailable then, are apparent now. Let us look at the situation in which we live. It is ironic that at the close of a millennium of unprecedented change and "development" which has brought us into the far depths of the heavens as well as into the inner spaces of the atom, we have such a vivid consciousness of mortality and death, and above all, the death of God.

While I hesitate to use words reminiscent of such a short-lived theology, still they express a vivid reality. In a very deep way as a people we are alone. One of the results of the Enlightenment and of the incredible achievements of science and technology is that we are alone in the world. Our own power, accomplishments, and sophistication have made us feel we are on our own. For us, success is not necessarily the blessing of God, nor are years of drought or disaster the anger or displeasure of a god. Unlike ancient or simpler civilizations, or even earlier generations, we are not sure for what we can turn to God. Good weather? The success of a meeting? Healing in illness? The miracle of a cancer cure? The protection and liberation of the oppressed? Safety for our children? Direction in life? An end to violence and drugs? Peace in the world?

---

renewal of Contemplative Life in the U.S. and the retrieval of the Christian mystical tradition will not be small.

5 See Constance FitzGerald, "Impasse and Dark Night," in *Living With Apocalypse*, ed. Tilden Edwards (San Francisco: Harper and Row, 1984). This is presently available with a companion essay, which develops the Dark Night further, "The Transformative Influence of Wisdom in John of the Cross," in *Women's Spirituality*, ed. Joann Wolski Conn (New York: Paulist Press, 1996). A more extensive study, "Transformation in Wisdom: the Subversive Character and Educative Power of Sophia in Contemplation," will appear in a collection of articles written by the members of the Carmelite Forum to be published by the Institute of Carmelite Studies, Washington, D.C.

God died in the concentration camps and the totalitarian dictatorships of this century, God was silenced in the Enlightenment and, according to Edward Schillebeeckx, with the disappearance of God, at least in the Western portion of the world, the individual person as a human subject also disappeared. God is dead, and as a consequence so is the human race.[6] The evidence that we have lost our humanity is all around us in organized terror, torture, ethnic cleansing or genocide, political murder, starvation on a mass scale, homelessness, increasing neglect/marginalization of the poor, and exclusion of the immigrant.

We see signs of death in racial hatred, escalating violence and abuse of women and children, in suicide machines and abortion, in drug addiction and drug sales, in the often lonely suffering of AIDS victims and the escalation of cancer deaths. We see signs of death in an alienation between women and men which shows itself in sexism, divorce, abandonment and rape on the societal level, and in the Roman Church in the absence of more and more women from the Eucharist, on the one hand, and in the inflexible and uncomprehending position of many Church leaders, on the other. The alienation women feel today and their struggle for equality and mutuality is just a part, a deep part, of the much broader alienation of the human species from the rest of the earth community.

When God is dead, not only is the human subject dead, but the cosmos itself is dead as either a subject or an object of respect. When humanity set itself up as an alternative to God, the human subject began to die and with it reverence for creation. Vaclav Havel, in addressing the situation of the world prior to the failure of communism in Eastern Europe, believes it was due not to East-West tensions, but to the spiritual condition of modern civilization. His devotion as a playwright to the theater of the absurd springs, in fact, from his concern for meaning, inasmuch as the theater of the absurd throws us into the question of meaning by manifesting its absence. He explains:

> As soon as man [sic] began considering himself the source of the highest meaning in the world and the measure of everything, the world began to lose its human dimension, and man began to lose control of it.[7]

6 See "Introduction to the International Congress for Theology" in *On the Threshold of the Third Millennium*, Concilium (Philadelphia: Trinity Press International, 1990), 14.

7 *Disturbing the Peace, Conversations with Karel Hvizdaia*, Knopf, 1990.

We see the destructive effects of modernity's "flight from the world" all around us. Numerous species become extinct every day; the rain forests are being destroyed along with the ozone layer that shields us from the rays of the sun. The air, the water, and the land are polluted. We recognize today the real possibility of the death of *the earth as we know it.* We have failed for a long time to understand our own place in the earth community and the absolutely essential connections between ourselves as the human species and all other species of life. In consequence, for the first time in the history of the cosmos, we face the possibility of the death of humanity as a species and the death of the earth as our home.

"Where have you hidden, Beloved, and left me to my moaning?" cries John of the Cross in the first lines of his beautiful, classic poem, *The Spiritual Canticle*. While this is the mystic's cry sounding through the ages, *it is our cry, the muted cry of our nation*. This is the cry of desire known by every person who has ever earnestly sought God. The desire for God is everywhere crying out! The miracle is that the contemplative cry of the people, of the whole earth community, is no longer silent and invisible, but rather prophetic and revolutionary. It rings through the universe and we must "not lose the thread of desire that from the depths of a disfigured world, groans with unspeakable groans to proclaim the birth of the new creation."[8]

Today the whole world cries out with the desire of the mystics, "Where have you hidden and left me to my moaning?" We hear on the one hand, the abandoned, the poor, the homeless, the dying, the elderly, the oppressed, the tortured, the martyred, the refugees, the rejected, the starving, the marginalized, and the abused, and on the other hand, we hear our nation, our people, the pleasure-sated and consumer-burdened, the addicts, those wracked by doubt, those disillusioned by government, those committed to justice and equality. All cry out. Even the scarred earth itself cries out and rages against its devastation. Creation groans with its desire and its dream.

All this has become a great cry of desire for life, freedom, and resurrection, a cry to the God of life who brings liberation out of every type of death, a cry for a new vision, a contemplative vision. This cry reaches beyond the collective of the dying communist systems and equally beyond the possessive individualism of capitalism. It is a cry for

8 Bingemer, 479.

recognition of the connectedness of everything in the cosmos, and consequently, a desire for contemplation and transformation, even though this is unnoticed for what it is by most people. The contemplatives realize it from one side; the theologians, led initially, perhaps, by some of the liberation theologians, realize it from another side, and both stand at a common meeting point trying to understand and articulate the time.

This universal cry, undermining the familiar dichotomy between contemplation and action, experience and theology, is what makes contemplation different today. The cry is within us. My own personal cry is overcome by, engulfed in, the cry of the earth, and it is within me. The desire for God becomes sheer passion when joined to the cry of the world. Religious people formerly approached God by turning aside from the world to some degree, but today we come with the world inside us; we are motivated in a very direct way by the earth, the people, the poor, the women and children. The relationship, or better the identification, is profound and it is experienced by many people.

We come, therefore, for the purification of the world, the transformation of human consciousness and human desire, and the completion of the image of Christ. We come as a last resort, in a sense, because we Americans have lost faith and hope in human power and human reason. We are driven to interiority, to contemplation, to the desire for the experience of God's love by the poor and oppressed, by, in fact, the image of God scarred in the world and on the face of the earth. We cannot continue to manage the world humanity has created with the skills, the minds, the wills, the memories or the imaginative paradigms we have. And it is the poor, the suffering earth, the pleasure sated that drive us in prayer toward contemplation, toward an understanding of our darkness and the need for the presence of contemplative vision, wisdom and love in the world, that is, to God's vision. Responsibility for the world, not just for individual well-being, moves us to contemplation.

This is the new "text" and this is the contemplative experience today. It rings with the ageless cry of the mystic to the Beloved:

> Where have you hidden?...
> Reveal your presence!...
> Extinguish these miseries...
> Who has the power to heal me? (SC 1, 6, 10, 11)[9]

[9] Most references to John of the Cross are inserted in the text. A=Ascent of

No gentle nor consoling cry is this as we sometimes imagine when we read John of the Cross' words of yearning and envision him swept away by a love far beyond our reach. Neither does this kind of yearning come at the beginning of the journey, but to those experienced and seasoned by life, often to those who have been generous and creative in using the resources available to them. For John, the mystic's cry of desire is not a disembodied experience, but always has a real life context in which there is a profound relationship between interiority and everyday existence.

REFASHIONING OUR VISION AND REFASHIONING THE WORLD

This explains why the mystics' painful cry for God in dark contemplation suggests a way of understanding our experience as North Americans. It offers some guidance for our future and some way of redirecting the movement of human desire toward God and refashioning our vision. Today we experience darkness in two distinct though related ways, like two sides of the same coin, that is, in our indissoluble connection or solidarity with the poor and marginalized, and in our own personal and societal failures.

First, the poor and the marginalized are our darkness, the darkness of humanity which is back-lighted for us today. In "the poor" our violence is unveiled, to use Gil Bailie's dramatic words. We know God's presence precisely because we see so clearly and so painfully the battered people and the scarred earth, an image of the Crucified One who is through time and space the darkness of humanity. In fact, in the poor people and the poor earth we recognize the way the image of Christ has been defaced by human desire.

Second, all around us the failures and limitations of our technological world and our philosophy are illuminated. We have rejoiced in the countless wonders of our technological age, in the magnificent exploration of the universe, the incredible life-saving medical discoveries, and the whole evolution of computers and cyberspace, but we seriously question our national ethos of unlimited development. We are proud of the

---

Mount Carmel, DN=Dark Night, SC=Spiritual Canticle, LF=Living Flame of Love. Quotations are from Kieran Kavanaugh and Otilio Rodriguez, *The Collected Works of St. John of the Cross* (Washington, D.C.: ICS Publications, 1991).

American desire to be free and self-reliant, to reach every frontier, to protect individual rights and provide opportunity for individual advancement, but more and more of us are confused today by our excessive, possessive individualism. Women particularly realize the destructive limitations of the autonomous self and the loss to society when the value of connection and relationship is muted and marginalized for the sake of pleasure, power, advancement, and success. We are burdened by our national pride that must be first at any cost. We recognize the exploitative, oppressive, selfish nature of many of our national social policies and our international relationships. We see the results of our need for comfort and pleasure in a consumer mentality that believes we have a disproportionate right to the world's wealth and resources regardless of the deprivation and mortal affliction it brings to poor, "undeveloped?" peoples, and notwithstanding the destruction of our mortally endangered earth and its eco-systems. Thus, as David Tracy asserts:

> The embrace of modern science, technology and industrialism... has helped to render present time for many an empty time—bereft of memory, free of hope, powerless to resist. The consumerism of our age is a relentless attack on the soul of every individual and every tradition.[10]

We experience desire gone awry and the failure of our national vision. As Robert Orstein and Paul Erhlich, authors of *New World, New Mind* explain, we cannot continue to manage the world humanity has created with the skills, the minds, the imaginative paradigms, the historical memories, or lack thereof, bequeathed to us by modernity.[11]

We feel *some* affinity, therefore, for the contemplative experience described by John of the Cross:

> At this stage persons suffer from sharp trials [and darkness] in the intellect, severe dryness and distress in the will, and from the burdensome knowledge of their own miseries in the memory, for their spiritual eye gives them a very clear picture of themselves (LF 1.20).

10 "On Naming the Present" in *On the Threshold of The Third Millennium* (Philadelphia: Trinity Press International, 1990), 69.

11 Robert Ornstein and Paul Ehrlich, *New World, New Mind, Moving Toward Conscious Evolution* (New York: Simon and Schuster Inc., 1989)

Our faith in the god of reason, progress, and finally technology, has left us without transcendence, without meaning, and without hope. The gods of modernity are dying, and with them our hope. In fact, one suspects some relationship between the experience of the mortality of humanity as a species and our society's pursuit of immediate pleasure, stimulation (drugs, sex, food) and unparalleled comfort with no thought for the effects on future generations. What kind of unconscious hopelessness drives those who do not even care about, much less provide for, the next generation or future generations?

John of the Cross' writings suggest that the clarity of our self-knowledge is in itself the embrace of the God we have silenced. Furthermore, opening ourselves to the dead-endedness and limitation of our national creativity unleashes a thirst for God that we will only feel when we come face to face with the deficiency of our present knowledge and the failure of all we have trusted as a people. When our history and tradition, that is, all we remember and count on for our self-understanding, are not sufficient to guarantee our future as a people and as a species, we need to recognize this emptiness as a yearning for God which is only possible when hope in our own abilities fails us. Is it conceivable for us to understand the breakdown of love, mutuality, and fidelity, the disappointment of human desire on so many levels, as a hunger for God that we will only sense profoundly when human desire is thwarted and betrayed? Inbedded within the experience of loss of meaning and imagination abides the possibility of wisdom and new vision; buried within the painful feeling of being abandoned and on our own lies the seed of a mutuality and fulfillment already in process.

In one way, it trivializes the writings of the mystics to suggest our national darkness has any relationship to John of the Cross' contemplative night, but in another way, looking at our national psyche within John's framework allows us to believe, to hope and to continue to desire. The challenge is to understand what is happening in order not to turn away from the faint light of contemplative wisdom and love that is showing us to ourselves.

> Now with the light and heat of the divine fire, [the soul] sees and feels those weaknesses and miseries which previously resided within it hidden and unfelt, just as the dampness of the log of wood was unknown until the fire applied to it made it sweat and smoke and

> sputter. And this is what the flame does to the imperfect soul (LF I.22).
>
> It is not gentle but afflictive... Neither is the flame refreshing and peaceful but it is consuming and contentious, making a person faint and suffer with self-knowledge (LF I.19).

This is actually a hopeful time, when theological faith, hope and love, *the gifts of God*, are being accessed. We are forced, as it were, to accept the alternative vision of faith, hope and love, and so pass over into the perspective of God. The question is this: can we receive the darkness of this *empty time and barren space*, personally and collectively, as the love and care of God in our lives desiring to purify our desire as persons and our national dream as a people? Can we hear God calling us to a more contemplative time when we will be able to see and appreciate a new vision, hear within ourselves a new voice, experience a new faith and love capable of creating new paradigms for living as a part of all life on earth and in the universe? We are challenged by the dead-endedness around us to mature to a different level of existence. The call for openness to God is radical!

The Need for a Quantum Leap

Ornstein and Erhlich believe that our species cannot evolve quickly enough biologically or even culturally to do this naturally. The human mind is failing to comprehend the world it has created.

> Human inventiveness has created problems because human judgment and humanity's ability to deal with the consequences of its creations lags behind its ability to create. There is now a mismatch between the human mind and the world people inhabit. The mismatch interferes with the relationships of human beings with each other and with their environment [and with all other species on the earth]... The serious and dangerous mismatch is this: civilization is threatened by changes taking place over decades, but changes occurring over decades are too slow for us to perceive readily... At the same time, the changes are much too rapid to allow the biological or cultural evolutionary processes to adapt people to them. We are out of joint with the times, our times.[12]

12 Orstein and Ehrlich, 10–11. I am completely dependent on their thought in the following paragraph.

Biologically we are still programmed for instant response and instant remedies. These are some of the default positions lodged in the human mind which Ornstein and Ehrlich say cannot be transformed or refashioned completely and certainly cannot be genuinely changed in our lifetime. Although we are evolving, our mental machinery will not change biologically in time to help us solve our problems. We cannot wait for the necessary tens of thousands of years until natural selection does its job in order to solve problems such as runaway population growth, the collapse of ecological systems and the approach of thermonuclear Armageddon. We must therefore find ways of replacing our old minds with new ones by means other than the evolutionary process. Orstein and Ehrlich suggest a new kind of education, namely, the creation of a process of conscious evolution. They insist we need to be literate in completely new disciplines.

I would like to suggest contemplative interiority as one of these "disciplines" and education for contemplation as essential to the new kind of education required, and I suggest it with profound seriousness to educators on every level. We do, indeed, need new minds, as well as new intuitions, new wills, and passionate new desires, since many of our highest achievements, Ornstein and Ehrlich explain, represent only a refinement of the old mind, not a new kind of perception.

Reflection on the sphere of the mind, according to Thomas Berry, the self-styled geologian, is imperative because the other powers of the earth seem now to have given over to the "mindsphere" the major share of steering the course of earth development. The earth that directed itself instinctively in its former phases seems now to be entering a phase of conscious decision through its human expression. This is the ultimate daring venture for the earth, Berry claims, this confiding its destiny to human decision, the bestowal upon the human community of the power of life and death over its basic systems.[13]

When Vaclav Havel some years ago addressed a joint meeting of the U.S. Congress, he, too, called for "a global revolution in the sphere of human consciousness." He proposed that the salvation of the human world lies nowhere else than in the human heart, in human meekness

13 See Thomas Berry, *Dream of the Earth* (San Francisco: Sierra Club Books, 1988), 19. Berry enumerates five major components of the earth's functioning: the geosphere, the hydrosphere, the atmosphere, the biosphere and the mindsphere.

and human responsibility. This belief in the importance of the transformation of individual consciousness for global change, reinforced by Havel's very life, is supported also in *Megatrends 2000*.

> It is the individual who changes himself or herself first before attempting to change society. Individuals today can leverage change far more effectively than most institutions.... movements (peace, environment, etc.) were built one consciousness at a time by an individual persuaded of the possibility of a new reality.[14]

Is the time of national darkness and national truth (dark contemplation or dark night), evidenced by the breakdown of structures and the awareness of national limitation by many, truly a portent of the kind of global revolution or transformation called for by Vaclav Havel, a call coming out of his own long dark night under Communism?

John of the Cross, in his commentary on the *Spiritual Canticle* poem, written while he himself was in prison, speaks of the time of union with God after the searing purifications of the Dark Night when the person takes on the mind of God, the desire of God, the will of God, and the memory of God, in other words, a radically different perspective. Most of us think of such fundamental change as next to impossible in this life. I would like, however, to suggest it as a viable option for the human community, one imbedded in and promised by our Christian tradition, one we cannot continue to undermine, make light of, excuse ourselves from or marginalize. We cannot afford to bypass contemplation, interiority, and desire for God, as though they were esoteric experiences for the lazy or unbalanced elite, but not for us who value above all else reason, sanity, and the ability to control our own destinies. Certainly without contemplative prayer and the transformation it really can effect, the deepest dimension of the human person and of humanity itself lies forever dormant and beyond our reach. But even more, without it the true evolutionary possibilities completely dependent on the inbuilt purpose and aspirations of the human soul are beyond us. I deeply believe this is the era of contemplation—Thomas Berry calls it the "mind-sphere"—and the stakes are very high.

We need to understand and to speak, therefore, of the unleashed power, influence and freedom of contemplative love and wisdom, of

14 John Naisbitt and Patricia Aburdene, *Megatrends 2000* (New York: William Morrow and Company, 1990).

their ability to pass beyond the limits by which both person and humanity are confined, the boundaries within which human consciousness, desire, culture, evolution and religion are now enclosed. Contemplation can bring within the realm of possibility the purified imagination able to create not only a global economy and *world* community that make a human life more possible for the poor, oppressed and marginalized of the world, but even the paradigm shifts and transformations required to invent a new kind of *earth* community where we reinhabit the earth in a truly human manner.

John of the Cross seems to imply that if we are to continue to grow and if the desire which constitutes our being is to reach fulfillment, the time will come when God's light will invade our lives and show us everything we have avoided seeing. Then will be manifest the confinement of our carefully constructed meanings, the limitations of our life projects, the fragility of the support systems or infrastructure on which we depend, the boundaries through which we shall never break, the dreams that will never reach fulfillment, the darkness in our own heart. Initially, this is a very dark time but it is also a contemplative experience when God's loving light embraces us with a power that is staggering both in its seeming destructiveness and in its potential for new vision, deeper love, radical transformation. The irony of the situation is that we experience this light as darkness, the nearness of God as unreachable and frightening transcendence, as no experience, in fact. The light is so excessive that we are blinded, we see and believe "nothing."

But this time of dark contemplation is, as has been indicated, an omen of radical revolution. For *the poor and oppressed* it indicates the process of liberation taking place as they become conscious of the desire so long repressed within themselves, let it emerge, release it as a cry of pain, and yet feel it at the same time to be energy for the struggle toward new life. For *North Americans* it marks a call by history and "development" to a contemplative or wisdom moment, a time of interiority, prayer and *waiting upon God* in a precisely theological faith and hope, that is, in a radical righting of the relationship with God and with one another. This means the cry indicates a time of painful knowledge and deep purification of national desire and resolve. It promises, however, the possibility of true liberation and transformation, a whole new view of what it means to be the *human* species on the planet earth.

All this is meant to show that the mystical tradition interpreted for today throws some light not only on our inner lives, but even on our life as a nation and on humanity in its present stage of development. We always speak as if the mystics' experience of God is the end of the life process, almost beyond our reach. I suspect their experience may be a beginning for us. The mystics not only give us paradigms from the past that enable us to interpret our present experience, but they also provide the materials to create new paradigms that unlock a way into the future and unveil the horizons to which contemplation, or the love-embrace of God's spirit, throws us open and makes us available as a people.

Signs of New Life, Transformation and a Cosmic Vision

In the midst of the signs of death, there are already signs of new life and new vision not accessible ten or fifteen years ago. We see them in the revolutions in the Philippines and Eastern Europe, unfulfilled or tragic as these seem at times, in the Havels and the Chinese students of Tinneman Square who have suffered the long dark night of the human spirit, in the energy of the poor who suffer the dark night of hunger, deprivation and abandonment and yet gather together in the basic ecclesial communities in faith and in hope, in the official dissolution of a policy of apartheid and the peaceful change to majority government in South Africa, in Oscar Romero, the Maryknoll Women, the Jesuits and their women companions, the Trappists of Algeria, a bishop like Frank Murphy, and so many unnamed others who rise up in death within the people with the promise of victory for the oppressed in a voice stronger than anything they had in life.

We see signs of new vision in the passing of dictatorships, in the repudiation of communism and the claiming of freedom, precarious as this freedom has proved itself to be. We see signs in the fragile movement toward peace in Northern Ireland, in the decision for unity in Germany, in the fragile collaborative efforts between Russia and the United States, in the faltering peace negotiations between Israel and the Palestinians, in the consciousness of the West when faced with the starving refugees of Africa, and in the effort to solve international disputes by diplomacy. We see signs in the strength and respectability of the feminist movement in theology and life, in the commitment to justice and equality, and in communities who live together in harmony, genuine love and fidelity thereby bearing witness to peace and mutuality in a world where so many live isolated, lonely, disconnected and uncommitted. We see signs of life and vision in the dedication to financial, economic and social responsibility against staggering odds, in the dialogue among world religions, and in the growing awareness of the condition of the earth. Lastly, we perceive signs of hope in theology's recognition of the ill effects of having ignored as integral to the Christian faith the mystical dimension, and the confinement of contemplation to an airtight compartment of the Church's life for centuries. We see signs of life in human awareness and life-giving service on so many levels.

The price for this new insight and compassionate love, however, seems to be darkness, suffering and even death. Our gods have to die before we reach for the God who is beyond all our human images and projections and who waits over the brink of the known in the darkness. The signs of new life appearing among us are somehow the other side of an emptiness that paradoxically is not only bringing us closer to God but also purifying our desire and imagination and moving them toward transcendence of what has been.

John of the Cross holds out to us the promise of union with a God beyond all our inadequate images and finite gods, a union on the other side of this dark and empty time. In fact, the dark emptiness is already the beginning of this communion. The same light and love that cause the painful self-knowledge and empty questioning effect the transformation. Love will, if we accept it, answer the world's cry of desire and overtake our lives with tender communion, mutuality and an entirely different perspective which we of ourselves cannot produce by sheer determination and reason. The time of this Dark Night is in the end a hermeneutic of not only our immature or long-lost images of God but also of the Enlightenment's autonomous self!

In the very last stanza of *The Living Flame of Love*, John of the Cross sings of this transformation as *the awakening of God:*

> How gently and tenderly
> *You wake in my heart*
> Where in secret you dwell alone
> And in your sweet breathing,
> Filled with good and glory
> How tenderly you swell my heart with love (LF III).

This awakening of God within the human heart is, of course, *human* awakening, but because such a radical conversion of perspective and affectivity is engendered by the silent tenderness and inspiration of love, it is experienced as God's waking up and breathing love from the very core of a person to every thought, emotion, desire, action. When God "moves" in this way, John explains, everything in the universe is experienced as connected and moving together in harmony. All things disclose the beauty of their being and the root of their life. In consequence, every form of life is known and valued in and through God, the ground, source and center of not only the human but of everything in the universe (SC 15.25, 36.6, 39.11). The contemplative, therefore, sings without apology:

> My Beloved, the Mountains
> and lonely wooded valleys
> strange islands
> and resounding rivers
> the whistling of love-stirring breezes
>
> the tranquil night
> at the time of the rising dawn,
> silent music,
> sounding solitude,
> the supper that refreshes, and deepens love (SC st 14–15).

Now everything becomes the Beloved one and, therefore, part of the human person who is united with God and who now experiences herself as one species connected to every other form of life on the earth (SC poem st. 14 and 14.4–9, 24.6). This identity with the entire cosmic order within the contemplative and the discovery of the earth as a *living* organism are the foundations for an intimate and compassionate human presence to the earth and to one another as humans. It opens up the capacity for listening to what the earth is telling us. John writes of an immense, powerful voice which sounds in the soul, the voice of all the wonder of God heard in the voice of creation (SC 39.8–9, LF IV.10–11).

> The soul becomes aware of Wisdom's wonderful harmony and sequence in the variety of her creatures and works. Each of them is endowed with a certain likeness to God and in its own way gives voice to what God is in it. So creatures will be for the soul a harmonious symphony of sublime music surpassing all concerts and melodies in the world (SC 14/15.25).

The contemplative person stands, as it were, with the Creator of the Universe who awaits us in the future and calls us to completion by the desire, the dream, implanted not only within human *being*, but also within the earth and all its species as an organic whole.

If we Americans admit to an experience of God at all, we usually experience God through earthly realities, through the beauty of creation and the wonder of human love. A child, for example, is able to believe in God's love when she has been loved by her mother. The desire for God grows and develops imperceptibly in the ordinary experiences of life. This is the way it is meant to be. But in the contemplative experience of God "awakening" and filling human desire, the whole of created reality, the whole cosmic organism, is experienced as a part of Infinite Being. One's basic perspective changes. One "has God's view of things" (LF I.32). The contemplative truly knows and sees everyone, everything, the whole earth, from the divine perspective and with the love of God and, therefore, with the desire of God for the world (LF 4.2–5).

John writes at length in many places of this transformation of all the powers of the person. The mind no longer understands with the vigor of its own natural light, but with the divine light. This is, in effect, the transformation of the mind through a new kind of knowledge, Holy Wisdom. The human capacity to love is changed because the mystic is so transformed by the unitive experience of God's love that she actually loves with the very love God has for us. Even the memory is changed in this union by the sure hope of a beckoning Future. John says the mind is God's mind, the will is God's will, the memory is God's memory and the person's delight and desire is God's delight and desire (LF II.34). All the powers and energies of the person move in love (SC 28.2–3, 8).

What is particularly significant is that for the contemplative person the old way of knowing and loving is gone. The process of transformation and conversion is really an irreversible evolution which amounts to a very radical re-education and transformation of consciousness and human desire that cannot be controlled by short-sighted human governments or fearful Church authorities (SC 26.13–17). This is the reason contemplation is so subversive and why, prior to the re-emergence of both popular and theological interest in it in the latter part of this century, it was so decisively muted for nearly three hundred years following the condemnation of quietism, as Joseph Chinnici so persuasively explains.[15] As long as contemplative transformation is looked upon solely as independent access to the divine, it will continue to be considered threatening rather than enlightening to the status quo. If, however, Ornstein and Erhlich are correct in suggesting the need for not just the refinement of the old mind, but rather for a quantum leap in which our old minds are replaced by new ones, then contemplation assumes an importance religion, government, educators and leaders in the Churches have long denied it.

Conclusion

One wonders if there is a way in our extremely complex, sophisticated culture that education for contemplation, understood as a process of intentional consciousness evolution, can ever be seen by our struggling, violence-racked society as a "new discipline" in which it is imperative that our people be literate. What if there were "schools" of contemplation or interiority all over the country, in business, in medicine, in education, in government? What if from pre-school to university *the art* of quiet reflection and listening, *the art* of prayer, *the art* of accessing the depth dimension of being human, without which we will never be fully human, were taught, nurtured and esteemed as central to all education? What if this manner of valuing the contemplative side of experience and development were seen as a response to and understanding of the darkness and fragmentation which afflict us as a nation? What if, in other words, the desire for God crying out among us were truly heard as we enter the twenty-first century?

I am suggesting that the call to contemplation be seen as integral to human self-understanding and as an absolute imperative of American education. Most suppose it is impossible for many people to become contemplative, as if this were an esoteric experience reserved for the few, a luxury for those who have leisure to be with God or simply consumer products for a bored elite. But in the Carmelite tradition, the prayer that leads to the divine embrace of contemplative communion and transformation is not an obscure discipline, nor a complicated exercise, but a relationship developed day by day through fidelity to a presence that per-

15 "The Politics of Mysticism: Church, State and the Carmelite Tradition," was presented in 1990 at *Carmel 200* in Baltimore.

vades every facet of our lives and whose image can be found etched in the length and breadth of the world.

The call is before us in both signs of death and signs of new life. We are summoned to recognize the shape of the desire for God asserting itself among us with more and more urgency. As North Americans we are greatly to be pitied if we cannot grasp our post-modern, post-enlightenment experience as an invitation to reliance on the God who truly is the mystery and energy pervading all things and who dwells darkly and secretly, as Karl Rahner writes, "in [the] nameless and pathless expanse of our consciousness" exposing to us the false visions of the present which afflict us while slowly revealing visions of hope for our future.

Meister Eckart *and* Dorothee Soelle *on* Suffering *and the* Experience *of* God

Gregory Baum

Several years ago, I received simultaneously two small books on the spiritual life that made me think. Reading one right after the other made me aware of a similarity between them, even though they represent different centuries and recommend different ways of life. The first book, written by the Belgian philosopher, Jean-Francois Malherbe, presented a conceptual analysis of Meister Eckart's preaching.[1] The second, authored by the German theologian, Dorothee Soelle, offered a series of essays dealing with God's absence and presence in today's troubled world.[2] In this essay I wish to explore the curious and unexpected affinity between Meister Eckart, the great fourteenth century mystic, and Dorothee Soelle, thinker, prophet, and activist of our own day. In my synopsis of Eckart's mystical theology I am following Malherbe's brilliant analysis.

Meister Eckart

According to Eckart's preaching, following the divine call begins with "radical detachment": detachment from friends and family, from the work we do and our activity in society, and from everything else we hold dear. What is called for is a total renunciation of self-centeredness. Detachment does not mean dropping these relations and activities altogether: it means, rather, becoming totally indifferent to them, recognizing that attachment to them is an obstacle to God's coming. A person defined by his or her relation to the social context Eckart calls "the outer man," in contrast to "the inner man," which refers to a person's essence, inner core, or true identity, to which the Godhead is graciously present. Since the outer man hides the inner man and will not allow the inner man to appear, it is radical renunciation alone that will allow "the breakthrough" of a person's true identity rooted in the Godhead.

1 Jean-Francois Malherbe, *Souffrir Dieu: La predication de Maitre Eckardt* (Paris: Cerf, 1992), 116 pages.

2 Dorothee Soelle, *Es muss doch mehr als alles geben: Nachdenken über Gott* (Hamburg: Hoffmann und Campe Verlag, 1992), 159 pages.

Eckart counsels detachment from community, from tradition, and even from God. What does he mean by this need to rid oneself of God? He distinguishes between "the outer God," which is the God mediated by the Christian community to which one belongs, and "the inner God," also called "Godhead," which is the true divinity, the hidden and incomprehensible mystery of love and freedom which creates the cosmos and grounds human existence. Without detachment from God there is no access to the Godhead who dwells in the inner man and desires to be recognized and loved in return.

Translating this into a modern idiom, one might say that the outer man is the person as defined by his or her culture. Social scientists recognize that our self-understanding is largely mediated by the people around us and the cultural symbols in which we are immersed. Our self-perception, or the knowledge of who we are, is socially communicated. It can be argued, for instance, that the omnipresent market system prompting each player to strain after the best deal produces in them an individualistic self-understanding. Eckart seems to say that our socially constructed self is inauthentic, a distortion of our hidden identity and an obstacle to life in the Spirit. Even the God preached by the Church is part of the problematic culture to which we belong and hence unreliable and misleading. Unless we inwardly take leave from our culture and try to dismantle the inauthentically constructed self, there is no entry into truth, no discovery of the true self, and no union with the Godhead.

This detachment, Eckart says, is very painful. What we love has to be abandoned. The disappointments, failures, and sufferings caused by life in the world must not be resisted but inwardly welcomed, since they help us to empty ourselves of false hopes and hidden attachments. When we fall ill, we must make an effort to become well again, but in doing so we must not regard the illness as a meaningless and unfortunate accident. Suffering then plays an important role in this, the first phase of the spiritual life.

The second phase is "the break-through." At a moment that always comes as a surprise, the emptiness readied by detachment is suddenly filled with divine grace, and the outer man gives way to the inner man. In a moment of transformation or conversion, a person transcends her culturally constructed self and uncovers her true identity. Here a person discovers her rootedness in the Godhead, recognizes that she has been invaded by God's love, and in turn finds herself deeply in love with God.

In the break-through, a person's love of God is not an effort, not an ideal to be striven for, but a spontaneous surrender. Everything a person does is now grounded in and carried by the love of God.

Eckart talks about the break-through as an experience with which he is familiar. He was not alone in this. Many mystics talk about a sudden change of consciousness, a radical conversion, an unexpected switch to a new phase of the spiritual life. In fourteenth century Germany, this mysticism was widely practiced, not only in the Dominican Order to which Eckart belonged, but also in other religious orders and communities of lay people.

The break-through experience is described by Eckart in a unitive language that some Christians have found somewhat troubling. Since Christian piety usually presents God and the believer as two, as Creator and creature, a Thou and an I, it is startling to hear that after the break-through the two are more correctly described as one. According to Eckart, the inauthentically constructed self (the outer man) experiences God as the over-against, and hence there are two; yet the authentically constructed self (the inner man) is so open to the Godhead, so grounded in it, and so filled with it, that they are one. Here personal consciousness is not experienced as facing God as another, but rather as participating in a transcendent consciousness which is divine.

Many mystics share this perception. They like to quote the Apostle Paul: "Not I, but Christ liveth in me." Yet because the self does not disappear altogether—neither in St. Paul nor in the mystics—they often use paradoxical language to describe their experience. Eckart made bold statements that got him into trouble with the Church's teaching authority. To reveal the gracious union between the inner man and the Godhead, Eckart sometimes said that as persons need God to be persons, so God needs persons to be God; or worse, that as my existence is nil if God does not exist, so God's existence is nil if I do not exist. Eckart believed that without being outrageous he would not be understood at all.

With the break-through, Eckart continues, everything changes. The world and one's human relations now appear in a new light. The person thus transformed is able to discover the divine presence in creation and the human community. The people and the activities from which the person, suffering great pain, detached herself, can now be embraced with love. For the inner man, engagement in the world is a good and holy thing, guided as it is by the love of God. Nor does the person who leaves for a time the peace of contemplation find that by being involved in the world she is being starved of God. For God present in the cosmos and human relations nourishes and delights those who have been opened to God by grace.

Eckart himself led a very active life. He held high positions in the Dominican Order. He was responsible for the well-being of the monasteries under his supervision, and, to visit his charges, he made endless journeys through the German lands on foot.

Even suffering is transformed by the break-through. According to Eckart's spiritual teaching, the inner man recognizes that in the sufferings he undergoes in this world, be they physical or mental, he is not alone: God suffers with him and bears the burden for him. The person grasped by God's love will neither flee from suffering nor complain of his lot because he knows that God loves him through his suffering, suffers with him, and rescues him from self-pity, discouragement, depression, and despair. Personal suffering is here participation in the divine suffering, that is to say, God's pain over the world's sinfulness and the misery and destruction this has caused.

Jean-Francois Malherbe has called his book on Eckart's preaching, *Suffering God*. Christians suffer God first in the painful detachment from everything they love; they suffer God secondly at the break-through in the sense that they are invaded by God while remaining passive; and they suffer God thirdly by experiencing their own pain as a participation in the divine suffering.

After the break-through, the third phase of the mystical way is "the transformed life," a life of contemplation and action. In contemplation the person remains with God in a state of surrender and experiences support of the inner man; but even acting in the world does not estrange the person from God because she encounters God in the world. Yet Eckart was keenly aware that the break-through is not a once-in-a-lifetime conversion. Even after this radical experience, Christians recognize that their conversion has not been deep enough or that they have been tempted anew by the outer man and fallen into sin. The transformed life remains, therefore, open to ever new conversions, sustaining and deepening the original break-through.

One of the excellent aspects of Malherbe's book is that it brings out the dialectic in Eckart's teaching on the spiritual life. Some interpreters of Eckart stress his teaching on detachment and on the world as obstacle to union with God, and then present him as an other-worldly mystic, an heir of Neo-Platonism. By contrast, other interpreters stress his teaching on the divine goodness communicating itself in creation and human creativity, and then present him as a passionate world-affirmer untypical of the Christian spiritual tradition. Malherbe convincingly shows that to understand Eckart correctly we must take into account both of these emphases: the total detachment demanded of "the outer man" and the world-openness produced in "the inner man." According to Eckart, the transformed life continues to pass through a series of conversions in which persons recognize deeper layers of self-seeking and suffer God's coming, recreating them in love.

Dorothee Soelle

Dorothee Soelle's primary reputation is as a writer in the prophetic rather than the mystical tradition. According to the typology of Max Weber, the prophet has a strong sense of God's over-and-aboveness: the prophet receives from on high the divine message, God's judgment of the world, and then faithfully proclaims it to the community. Yet in the book under discussion, a collection of essays and reflections, Soelle reveals another side of her religious self-understanding.

At the beginning Soelle explains that until now she has avoided speaking of God in her teaching because she had found it impossible to find an appropriate discourse. Whatever one says about God is bound to be misunderstood since the term "god" has become part of the vocabulary of our religious and secular culture. In the churches, God is the Omnipotent One, the Almighty Father, the provident Lord of history who rules the world from above, founding and blessing the hierarchies in church and society that keep order here on earth, including the superiority of men over women. The image of God as ruler or heavenly boss gives rise to an ambiguous religious imagination, even if this lord is pictured as kind and merciful. For then people expect him to intervene from above to overcome the evil conditions under which they suffer, and do not intend to join the struggle for justice themselves. And if this lord does not intervene, people attribute a hidden divine purpose for the greater good of humanity to evil historical happenings, including catastrophic events such as Auschwitz and Hiroshima. Here the merciful

God becomes a monster. It is no accident, Soelle argues, that nations attach the name of God to their armies or put it on their dollar bills. She quotes with approval Meister Eckart's saying that to live a spiritual life we must be delivered from God. This is a reference, we recall, to Eckart's "outer God."

But is there another divinity? Yes, says Soelle: God is ever present in the world, even in today's world, despite its secularity. This is how she puts it: "I do not believe that people today experience God less than they did in earlier times. God's presence and God's absence are felt by us in rejoicing and in despair, and sometimes in the puzzling intermingling of the two. Life itself is deeply penetrated by the quality we call God so that we cannot avoid feeding on it and hungering for it. Only we do not realize this because we have been made speechless."[3]

We have been made speechless, Soelle argues, by the discourse of the churches and their theologians who in their maleness have protected the lord. They have imprisoned God in their Bible and liturgy, instead of allowing the Bible and liturgy to become the lenses through which we read our everyday experience. For it is there, in our everyday experience, that God is present to us.

To explain what this divine presence is, Soelle tells stories of people she has met and stories taken from her own life. She relies on stories, not on concepts. Concepts make God into an object of the mind, separating the known from the knower, and thus making God into an over-against. By contrast, stories are able to reveal God's in-and-throughness. The stories Soelle tells, many of which are taken from her experiences in the Third World, deal with the conversion to hope and love or the perseverance in hope and love under conditions that undermine hope and foster indifference. They tell of people broken by life who do not despair but move forward; people in great need who find the generosity to extend a helping hand to their neighbor; people making sacrifices for children, their own and others; people caught in an economy that excludes them who do not give up the vision of a just society. Experiences such as these are not the result of human will power: they contain moments of despair, the panic of helplessness, the surprise of empowerment, and a sense of being inwardly remade. Soelle argues that because people lack an appropriate language for these experiences of the *praesentia Dei* in

3 Soelle, p. 49.

the midst of their dereliction, they remain mute unable to express themselves. Without the ability to communicate, people remain untransformed by their experiences.

If we want to speak of the *praesentia Dei*, we have to appeal to people's memories. For there must have been in their lives moments of despair and rescue, or conversions from hatred to forgiveness, or fidelity to truth in the face of opposition, or other experiences of death and resurrection. Soelle also tells counter-stories of people she has met, who have no such memories, and who either long for power or revenge or are caught in cynicism and narrow self-concern. When medieval authors spoke of *Gotteshass* (hatred of God), she writes, this is what they must have had in mind. Since for her the crucial human option is between God and idols, she thinks that the sinister human beings, who remember no experience of redeeming hope or love, must have fallen prey to some idolatry.

Soelle, we notice, prefers stories to metaphysics. Yet her account of the divine presence in people's lives actually corresponds to Karl Rahner's theology of divine grace. According to Rahner, revealed in Jesus Christ is the idead that God's gracious self-communication in people's lives is universal—universal as offer, not necessarily universal as received—and that, therefore, wherever people struggle over matters of truth, love, and justice, they are not left to their own broken resources but experience in their lives a summons and a strength not of their own making. Rahner called this the divine "always already." Speaking to people of the Gospel is not introducing them to something that is foreign to them, since God has "always already" been present in their lives. Rahner's divinity—just as Soelle's—is a God from below.

Soelle suggests that her theology of divine immanence is in line with the mystical tradition. She quotes Eckart several times. Yet this theology was not confined to the mystics—a point I cannot develop in this essay. The theology of immanence is much more widely represented in the Christian tradition of antiquity and the middle ages, even if the theologians did not succeed in applying it consistently to their understanding of God's lordship.

Dorothee Soelle, we note, does not wish to abolish the biblical titles of Father and Lord. She shows that in the Old Testament, God as father usually referred not to the origin of humankind, but to the future destiny of God's people reconciled. And in the New Testament, God as lord was related to God's coming kingdom, and hence referred not so much to the present as to the future. If the titles of God's sovereignty are indeed eschatological, then their meaning in the struggle of everyday life is quite different: they do not legitimate earthly rulers and hierarchies but, on the contrary, question and delegitimate all unjust structures. That is why Soelle believes that the biblical message of justice is capable of rescuing the Bible from the patriarchal values reflected in it.

Where Soelle differs from Eckart and, I think, from Rahner, is in her emphasis on eschatology. God for her is not so much "ground" as "destiny." Eckart did not understand his mystical teaching as a theology of history. By contrast, Soelle relates the *presentia Dei* to people's willingness to suffer with the victims of history and struggle with them for a just society. If I may use a sentence I have employed in my own writing, Soelle has a profound recognition that Christian love in an unjust world transforms itself into a longing for justice and an impulse to act so that the heavy burden can be lifted from the shoulders of the victims.

If Eckart's mystical teaching can be summed up as "suffering God," so can Soelle's political spirituality. She writes, "Ohne Schmerz ist Gott nicht gegenwärtig."[4] She is fully aware that a religion that consoles people in their suffering and asks them to accept it patiently is a piety that has reactionary political consequences. This was, after all, Marx's principal objection to Christianity. Yet, while Soelle acknowledges Marx's analysis, she transcends it in her own original way. Soelle distinguishes three Christian approaches to human suffering. The first one, related to God as omnipotent heavenly ruler, holds that the suffering inflicted upon people is part of a benevolent divine providence that either punishes them for their transgressions or plays a hidden and unrecognizable role in bringing about a greater good. This is the classical approach according to which suffering must be borne with patience. The second, more modern approach holds that God rescues us from suffering, and that if we have great faith, God will heal our illness, remove our anxiety, enable us to succeed and lead us to personal happiness. This approach corresponds to the modern, individualistic, middle-class consciousness, often found in an extreme form in evangelical preaching on television.

A third approach, entertained by Soelle, does not counsel that we flee from suffering. Why not? Because God suffers with us and in us. We suffer from the brokenness of the world which damages our own

4 Soelle, 91.

flesh and from sickness and old age, the signs of our mortality. More than that, we grieve over the suffering imposed upon others, especially the massive suffering inflicted on people by the conditions of oppression. We lament the dreadful events visited upon humanity in the present. Following the option for the poor, we extend our solidarity to the powerless, helpless, and outcasts; we commiserate with them, feel some of their pain, and mourn our incapacity to help them. More than that, we mourn that the economic and political institutions to which we belong cause unemployment at home and massive hunger abroad, while we ourselves may even be deriving benefits from them. Following an observation written by Dietrich Bonhoeffer in prison, Soelle holds that in this suffering God suffers with us and in us.

Lest this teaching encourage any form of self-indulgence, self-pity, or masochism, Soelle distinguishes with St. Paul (2 Corinthians 7:10) between "worldly suffering" leading to spiritual death and "godly suffering" leading to conversion and salvation. Godly suffering, Soelle writes, is an expression of love that moves to action. Suffering has thus a political meaning which Marx did not suspect. By participating in God's suffering we receive strength and a new outlook that make us join the struggle for a more just and more humane society.

Saying that God suffers is not traditional Christian teaching. At one time the teaching that God the Father suffered was actually condemned as heresy. Critics say that the refusal to think of the deity as suffering reflects the influence of static Greek metaphysical thinking on Christian theology. I am not sure whether this is true. In any case, more recently, wrestling with the new darkness of the twentieth century, several theologians (Dietrich Bonhoeffer, Jürgen Moltmann, and Douglas Hall among them) have begun to speak of God's suffering.

Suffering, it seems to me, may be said to have two meanings. We suffer because we are personally vulnerable: we can lose our mind, our health, our family, our friends, our livelihood, etc. Suffering is here due to the lack of something which belongs to our integrity. One may call this suffering *ex carentia* (by deprivation). But we also suffer because our beloved friends are vulnerable: our heart is torn because they have lost something that belongs to their integrity. We suffer with them because we have extended to them our love and our solidarity; we suffer with them because we have reached beyond ourselves and identified ourselves with them; we suffer because an inner wealth or fullness has allowed us to give ourselves away. I call this suffering *ex abundantia* (out of fullness). Usually it is called compassion. With the help of this distinction we are able to speak of God's suffering while remaining faithful to the classical tradition. God cannot suffer *ex carentia*: God cannot lose what pertains to God's integrity. But God suffers *ex abundantia*: out of the divine plenitude God loves those who suffer, shares their pain, and bears their burden with them. This, it seems to me, is the traditional teaching that God is compassionate.

What is remarkable, therefore, is not that Soelle insists on the suffering of God, but that she believes that in our own suffering we "participate" in God's suffering. Here she clearly moves from prophetic to mystical discourse. On the whole, Protestants have shied away from saying that believers participate in the divine life: because we are creatures and because we are sinners, there is an unbridgeable distance between us and the Godhead. Moving beyond this point of view, Soelle holds that being touched by God and enabled to be in solidarity with the poor and the weak transforms people so that their suffering actually shares in God's own pain. Does Soelle believe that we can also be enabled to share in God's love and God's joy? To answer this question, we will have to read her subsequent books.

SIMILARITY AND DIFFERENCE

It is surprising that Meister Eckart and Dorothee Soelle have so much in common. Both recognize that the dominant culture produces a false consciousness. Both reject the authenticity of "the outer man" and the reality of "the outer God." Both acknowledge a need for a divinely initiated conversion that frees us from the prison of our culture, delivers us from the conventional image of God, and relates us to the true divinity, the God beyond God, God's transcendent immanence deeply implicated in our personal lives. For both Eckart and Soelle, it is important to move beyond pietism, beyond the master/servant and even the father/child relation to God, even beyond the divine "I-thou." For God is never simply an over-against but always and primarily an in-and-through. Both assign special meaning to suffering, i.e., "godly suffering" as opposed to "worldly suffering," and both speak of our participation in God's suffering.

Yet for Eckart God is primarily Alpha, while for Soelle God is primarily Omega. For Eckart, God is the origin and hidden ground of being who calls and empowers humans to rest in and act out of the divine life, while for Soelle God is the future reconciliation and pacification of history who calls and empowers humans to yearn for their destiny and work toward it.

Of course, the difference between these two Christian thinkers is even greater. To understand this difference, I wish to offer a contextual interpretation of their spiritual teaching.

Meister Eckart preached in the feudal order of fourteenth century Germany where people were profoundly identified with their family, their village, their parish, and their tribe, and where their mind-sets was shaped by the cultural and religious tradition to which they belonged. People were inevitably conformists: they accepted their inherited location in society. But in many of the towns, a new, more independent consciousness was emerging among the enterprising merchants whose intention to prosper and expand made them affirm the world in a new way and whose interests and travels took them far beyond the traditional boundaries. They wanted to act in society and, if need be, change it: they loved risk and adventure. Through their activities, they acquired a new sense of personal freedom and identity.

In his sermons, Eckart denounced the materialism and the profit motive that inspired the merchants. He believed they were moving in a direction dangerous to themselves and to society. Yet he did not take a conservative position: he did not call them back to the conformity of the village and the parish. Eckart, as we saw above, preached detachment from the tribal world. He urged people to detach themselves from what had been handed on to them by tradition, including the inherited religious teaching. Like the merchants, Eckart loved adventure and wanted freedom and action. Yet, against the merchants who sought their own advantage and were driven by selfishness, he proclaimed that selflessness and surrender to God offered an entry into freedom, world-affirmation, and action. If people were willing to empty themselves of their feelings and desires, he believed, God would invade them, convert their consciousness and reconstruct their personal identity. Then, united to the divine, they would discover the freedom to love creation, act in the world, and live adventurous lives. Here God, not selfishness, guided people from tribal solidarity to personal freedom.

What Eckart preached in his day was "a third way." He rejected the old way of conformity, saving only the best of it, namely, the quest of God, and he rejected the new way of materialism and self-promotion, while embracing the best of it, namely, the search for freedom. The third way he proposed was total surrender to God as surprising entry into freedom.

In his century Eckart was by no means alone in his search for a third way. There were other mystical preachers and teachers who shared his spirit. This, or at least a similar spirit, was found among the Dominicans and other mendicant orders who expressed their ideal in the motto "*contemplata tradere*." They opted for a third way which they named "*vita apostolica*," combining and yet surpassing the traditional "*vita contemplativa*" and "*vita activa*." The inner life was here seen as providing a new vitality that allowed these committed people to step out into the world and communicate to others what they had received. According to many authors, this type of spirit influenced Martin Luther in his proclamation of freedom.

What makes Eckart's preaching incomprehensible today is his emphasis on detachment. We are no longer embedded in a collectivist tradition from which we have to be liberated by detachment. Since we live in an individualistic culture, we are far too detached; we hardly know any more what tradition, community, and solidarity are. The market economy makes us look upon life as a game where each looks for his or her own advantage. But because we are so unattached, we also feel isolated, deprived, fragmented, and separated from the Spirit. This is the cultural situation in which Dorothee Soelle communicates her spiritual way.

Instead of detachment, Soelle calls for solidarity. Self-emptying for her means bracketing one's own concerns so as to embrace the concerns of the victims of society and the wretched of the world. Opting for the poor, bearing the burden with them, and supporting them in their struggle, rescues us from our isolated and fragmented state, allows us to see through the ideological dimension of our culture, and brings us in contact with God's own life, especially by participating in God's suffering.

In modern times, a good many people have reacted against the dominant culture embued with individualism, and extended their solidarity to classes or communities struggling for social justice. I wish to call this "the second way." Here people transcend themselves, live for others, yearn for emancipation, and make personal sacrifices to transform society. We often refer to these people as "the left." In a sense, Soelle belongs to the left. Yet it is also possible to say that Soelle's spiritual teaching offers "a third way." For if I read her correctly, she laments that the reaction of the left has only too often been associated with a single cause, the cause of a nation deprived of autonomy or of a class or sector of society subjected to oppression or discrimination. While such conversions to solidarity have saved people from isolation, the single-cause struggle in which they engage has created new injustices and cut new wounds. To seek rescue from self-involvement is indeed a step toward human salvation, but to join a struggle defined by a single form of oppression is a problematic enterprise guided by blind solidarity. According to Soelle's "third way," the entry into solidarity must pass through a divine break-through, making us ready to suffer with all human victims and thus correct the just cause we are serving in the light of all the injustices inflicted upon people. This conversion delivers solidarity movements from the one-sidedness, stubbornness and narrowness which often plague them. This conversion, according to Soelle, also implicates us in mourning and a share in God's own suffering.

Since this first book on the mystical dimension of the gospel, Dorothee Soelle has explored this theme in several of her publications.

Questions *from the* Earth Story *for the* Christian Story *on* Death *and* Ressurection

Moni McIntyre

For all life longs for the Last Day
And there's no man but cocks his ear
To know when Michael's trumpet cries
That flesh and bone may disappear,
And souls as if they were but sighs,
And there be nothing but God left;
But I alone being blessed keep
Like some old rabbit to my cleft
And wait Him in a drunken sleep.

From "The Hour Before Dawn"
by W. B. Yeats

What happens when the earth calls back its own in death? Who or what is doing the calling? How do Christians grapple with the mystery and reality of life and death? In this essay I wish to explore the phenomenon of death from the perspective of the earth insofar as this is possible and to see what challenges this poses for the church today.

The Earth Story

We are born into this life we now experience and are called out of it by some mystery that has both baffled and intrigued the most thoughtful and technically proficient persons the world has ever known. Neither philosophy nor science has provided satisfactory answers to our questions concerning the origin and destiny of the human beings we call ourselves and know so well, yet so little.

Prior to the life that we now perceive, there was a material existence. Our particles were present and interacting with the rest of existence. Until somewhat recently in whole earth time, we have been unconsciously differentiated bits of the cosmos. We have always been part of all that is. Genetically encoded to become conscious at a particular moment, our bodies have been in some form since the beginning of the material world. All that we are comes from what has come before us.

We were not created out of nothing at the moment of conception. The conceptus we became had been components of other entities long before Moses and Sarah, Confucius and our parents. The material awaiting our conscious expression permeated our world in all its splendor for eons before our appearance as human life.

Before our own conception, we were unaware of the material in the universe and that we were part of it and formed new lives from it. What was material and apparently passive was really incipient consciousness. Although we could not know that we would come to be, others were aware of aspects of the materiality that would eventually become the "I" of myself.

The seeds of our consciousness were also sown in the beginning. Our conscious selves have existed *in potentia* for all time. The material that carried our bodies also carried our consciousness. The configuration that we call "I" has taken all time to unfold. We could not make conscious contact with this material ourselves since the time for the "I" that describes the self had not yet come. The "I," the conscious self, waited for the moment in materiality that would allow the self to speak with conscious self-reflection.

Human beings tend to assign special status to particular moments of development. For example, even though the egg and sperm prior to fertilization were alive, we typically call the moment of conception the beginning of life. At this instant a unique pattern of cell division begins that will, if able to progress without untoward interruption, become recognizable as a person. At least equally significant is the fact that the ingredients leading to the formation of the particular sperm and egg have been waiting for all time to present themselves in their present form. These tiny fragments have cooperated in some way with other bits of cosmic marl to engender the lives we would eventually call ours.

That the ingredients of a future human life are as near as the air we breathe and the water we drink is an astonishing thought. These ingredients are within us as well. No other material for life forms exist. We are in the midst of the kaleidoscopic possibilities of all that is. The heart of the earth contains the profundity of our psyche. The vast and incomprehensible universe we have come from and know is all there is and all there has been and will be. In the whirl and mix that is the earth, we somehow emerge as conscious human beings who experience the intricacies of the cosmos in a myriad different ways in somewhat predictable patterns.

A young engaged couple can symbolize the multifarious possibilities for the earth. Particular particles of the universe have converged in human form and consciousness. At this moment they are poised to forge new careers and to contemplate their unique repetition of the exquisite and bloody moment of birth. In their presence we can imagine what has been as well as what will be. Humans in history have been mating and building since their beginning. Inevitably this engaged couple will surprise not only themselves but all who know and love them as they make their fairly predictable decisions for the first time in the history of humanity. Their contributions to humankind will likely include cultural enrichment and members of the next generation.

It is not enough for us that we exist. Human beings long for meaning, an ultimate meaning of the universe. We long to find ourselves within ultimate meaning. During one's lifetime, vital contacts with others help all parties to establish meaning in their own lives. While no one is capable of revealing fully the identity and complexity of the mystery that has first called life into being, other members of the earth community can help one to glimpse a whole amid the fragments that we can perceive. The group that forms community may discern what eludes the perception of the individual. Those who bond deeply with others develop an intensity that transcends their interpersonal encounters. While all conscious persons may be capable of envisioning another who is not physically present, those who bond may even imagine themselves connected in thought and intention with absent meaningful companions. A necessarily precarious network of significant relationships sustains our lives and gives us hope. Until death.

Death changes everything. At the moment of death we are aware of the absence of consciousness from the material we called "Marion" or "Anna." A disintegration occurs that leaves the visible materials in what looks to be an emptied state. Within a relatively short time, interaction with the tangible remains becomes meaningless and counterproductive, while attempts to contact the conscious part of the person are fraught with frustration and uncertainty. As long as body and spirit were unquestionably present, their lives could be closely and obviously linked with ours. In death a clear separation occurs. The body remains without a map for where the spirit has gone. The consequence of this detachment for those remaining may be one of unspeakable grief and unquenchable thirst for information and comfort. What held the deceased person together was the tension that is body and spirit. The absence of either is no longer life as we have known it.

What we think of as human life is the experience of the tensive reality that takes place when what was thought to be inert reveals the presence of the spirit of the universe in a new conscious life form we call person. Birth and death, as well as other poignant moments between them, occasion recognition of the incomparable value of particular expressions of perceptible but intangible energy. At these times, life in its cosmic magnificence may be apprehended in some proximate way.

No living person knows for sure what happens in death. We do know, however, that humans tend to give greater emphasis to the spirit that was embodied in conscious form. The relationship of matter and spirit undergoes a radical change. The tension between matter and spirit is released. The person before us changes form. Matter gives way to the spirit. The material components crumble into other substances and lie in wait to become other life forms.

After death, the spirit must also be somewhere near. If incipient consciousness is present before birth, mature consciousness cannot be nowhere and nothing. The intangible substance which leaves the body during the process of death must be present after death in some manner, however obscure and reconstituted. In death, that same spirit must remain with us if the world we know is all there is, but the spirit is no longer available to us in the form to which we had grown accustomed. The entities that were Marion and Anna, then, are not destroyed; rather, they are transformed in such a way that they are beyond our comprehension. We have hope that their present existence will become clear to us when we finally join them in our own reconstituted structure.

In the current of life, nothing ever stays the same. Even when we think it does, we are mistaken. The process of all life is the same: continuous connections of matter and spirit, both delights and disasters in their fusion, and disconnections of matter from spirit. As far as we know, it has always been this way, and it will remain this way until the end of time and space. Neither connections nor disconnections are ever total and never can be. As we continue to live, the world continues to be reconstituted. Our challenge is to reconstitute meaning continually since, of all earth dwellers, we seem to be the only ones at this point capable of articulating the notion of meaning for existence.

At present we are limited to this dimension of life. We cannot recall a previous existence because it is not perceptible to us in this material formation. Similarly, we cannot yet reach the dimension to follow. We may feel Simeon's kind of intense joy as we celebrate a birth, and we also may feel something of Mary Magdalen's searing pain as we weep at the tomb of a dear friend. Thougthful persons always walk away from such poignant moments wondering.

The Christian Story

Our planet has given rise to several prominent as well as obscure myths concerning the meaning of human life. In this section I would like to situate the Christian story within the context of the earth story and consider our need as Christians to appreciate the relationship between ourselves and the rest of the earth.

The Christian faith is one of the principal expressions of the earth's story.[1] Christianity is a particular historical and knowable path through which one may ponder and even encounter the mystery of the universe.[2] In the first immediacy, the Christian story provides an explanation for the undying hope that is within us. On their best days, Christians wait with hope and have faith in promises that the God who made us will not ultimately fail us. Christianity discovers fresh reasons for hope in a benevolent Other who loves with passionate identification and longing.

The Christian story assures us that there is nothing finally inexplicable. That we and every other member of creation are known, loved, and appreciated for ourselves. That someone just as human as we are has finally understood and communed with the mystery of the universe. That he has made sense of apparently senseless suffering and death. The Christian faith tells us that we are never alone. That just because we think we are the most evolved species that the earth has produced, the totality of meaning is not confined to our individual or collective intellects. We do not finally have to formulate meaning for all.

1 There are as many expressions of Christian faith as there are Christians. For purposes of this article, however, it is assumed that Christians are distinguished by their belief, in some fashion, in the death and resurrection of Jesus Christ.

2 "Christianity" as it is used here refers to the way Christians have lived out their beliefs in communities called churches.

Our belief in the Other, whom we call God, gives us an opportunity for relationship with the mystery of the universe. We experience God in the deepest vaults of our own mystery. Christians hold in common the expression of this mystery as three persons in one, and we name and grapple with them as lovers, friends, and members of one body. Access to God is limited only by our capacity to perceive mystery in all its fullness.

Christians come together in various kinds of communities of related minds and hearts to offer praise and thanksgiving, to celebrate life, death, and resurrection, and to challenge and respond to the challenges of our sacred writings. These assemblies, or churches, have always proclaimed the resurrection of Jesus, but they have never been able to articulate clearly and unambiguously what this means. They argue that God goes before us and accompanies us into the next dimension. Without God, they claim, nothing makes sense. Weeping at the fresh grave of a loved one, the Christian knows of nowhere else to go.

Ambiguities pervade the world in which we live. Questions of life and death haunt every thinking being as one tragedy after another presents itself. Although we may not defer the questions to another age, the collected wisdom of the ages ought not choke the answers of those who also have a right to breathe their own answers freely. Christian communities must ask and help others to ask the hard questions of every age.

Inevitably, Christianity assumed the shapes of the imaginations of dominant members of its sects throughout history. Voices long absent from the rolls of policy makers have strained to be heard in every generation. Despite the uneven treatment of the governed by both the governed and governors, as well as attempts by outsiders to squelch the spread of Christianity, believers in the Christian story have clung to the core of the myth. Tales of the courage of the truly faithful inveigh against the complicity of today's Christians who welcome injustice and celebrate it in the name of God and country.

Churches are recognizable forums for reflection on the identity of God and the earth. They proclaim and discover again and again that love is the way to God because God is love. The love of Jesus for the creation of God inspires and enables churches to engage in relentless struggles to love and serve the voiceless, homeless, and otherwise bound members of this world regardless of their success. With courage,

Christians must live the answers to the question of how to love so much of a world that seems bent upon denying its most basic need.

Both a study of the earth and the Christian faith as part of the earth assure us, in the words of Paul to the Romans, that ultimately "all things work together unto good" for those willing to trust and love despite their unfulfilled yen for answers. Churches must help us to appreciate the goodness of God expressed in creation. The Christian story has arisen from the earth, from the creation of God who created the earth. Its authentic components are congruent with the rest of the earth because it is in harmony with the creator of that earth. Listening to the story of the earth can lead us to the divine presence who is always present among us. We are, after all, of the earth.

We would do well to be open to learning from all members of the earth community. The life and death questions of every age must be approached with courage. As Christians, we strain our answers to these questions through the paschal mystery. We believe that God goes before us and is with us still. The responsibility for final answers is not ours. This helps us to perceive the burdens as light and the blessings as heavy.

Courage *and* Wisdom *for the* Ecclesial Work *of* Inculturation

Carl F. Starkloff, SJ

In October of 1993, at Anishinabe Spiritual Center in north-central Ontario, there took place a meeting called by its Ojibway name *Wassean Danda* ("Let there be light"). Some thirty Jesuits involved in ministry among indigenous peoples in different regions around the world met, along with the Society's general superior, Peter-Hans Kolvenbach, to examine this ministry, perhaps for the first time as an international endeavor. Drawing its inspiration largely from the experience and reflections of the Jesuits' thirty-second and thirty-third General Congregations, the group spent six days in intensive study and reflection on the meaning and "praxis" of inculturation. Central to the entire meeting were the elements that I shall discuss in this essay: burden, courage, and wisdom in the Church's work of inculturation of the gospel.

On the fifth and sixth days of the meeting, the group experienced two different faces of one aboriginal culture vis-a-vis the Society of Jesus and its mission. One could look into both faces and hear from the voices behind those faces many things that call for both courage and wisdom. On the afternoon of the fifth day, the Jesuits traveled over to the town of the "Three Fires People," the Ojibway, Pottowatomi, and Odawa: Wikwemikong on Manitoulin Island in Georgian Bay, to be hosted in a gala celebration by the members of Holy Cross parish. There was, first, a formal visit to the parish cemetery, followed by a eucharist presided over by Bishop Bernard Papin, with a homily by the Jesuit General, and several speeches of welcome by parish members. The symbolism there was a mingling of native blessings and traditional Christian hymns in the native language led by the parish choir. One native priest, two native deacons, and one native "DOW" (Diocesan Order of Women) assisted at the altar. Following the liturgy, there was an outbreak of picture-taking, then a visit to the band office to view paintings by native artists and early missionaries, and finally a tour of the reserve. That evening there was a lavish feast in the village arena, followed by more speeches and an exchange of gifts between the Jesuits and tribal members. The evening concluded with a two-hour exhibition of tribal dancing and local interpretations of the traditional "clog dancing" that had been introduced from Europe into the Anishinabek culture.

Although it was a joyous event, warm and welcoming, there was a quiet undertone to the party that called for wisdom to interpret. The General's homily was a key to this, featuring as it did, among many positive aspects of mission history, an avowal that many mistakes had also been made by missionaries, and a request for forgiveness and understanding. Everyone present knew that euphoria should not be the only emotion for the occasion. The realism of this wisdom became very evident the next day, when the group traveled to the nearby University of Sudbury for lectures, visiting, another art exhibit, a feast of native cuisine, more speeches, and a native dancing exhibition. One of the native responses to the address by a Jesuit historian continued the very "Catholic" tone of the previous day, with much nostalgia and praise for the Jesuits. But the second response, by the chairperson of the native studies department, made it clear that "celebration" was not the tone of his message. Although he expressed some appreciation for the support by the University's Jesuit administration of the native studies program, he proposed a "hermeneutic of suspicion" to be applied to the Jesuit's speech, and argued that an Anishinabe historian would read that history much differently. In his second speech later in the evening, he argued that Christianity is a foreign religion for the native people, and that they could be true to their own identity only by following their own "good red road." Christianity could never truly satisfy the spirit of the native people, as was demonstrated by a citation of even current actions by the Church that continue to demean native culture.

It was a sobering experience, though not at all a surprising one for the Jesuits in attendance, who had been examining such problems as these for the previous five days. Everyone knew that wisdom (with its more prosaic quality "prudence") must be the animating spirit of their work, and that a certain fund of quiet courage must accompany this. Perhaps what is even more necessary, one might suggest, is what one of our number called, using an adaptation of the language of Ignatius Loyola, "the Fourth Degree of Humility." That is, those white ministers now engaged in work with native people must be ready to deal with forms of "opprobrium," with accusations and denunciations. But unlike Ignatius' Third Degree of Humility, at least from the viewpoint of native people, they cannot claim to "have given no cause for it." While they may wish to consider themselves to be of a more "enlightened" mentality on native culture than were their predecessors, they are still the heirs of those predecessors and share in the oppression that is seen to have resulted from earlier missionary activity.

This brief narrative, therefore, leads me to include in my essay an expression of the need, not simply for *courage* to do inculturation, but for the gift of *wisdom* to sustain the task. Inculturation may indeed include dramatic and startling events, perhaps even violent death in some quarters of the world, but the commonplace manifestation of courage must be the virtue of patient wisdom to help us "hang in" for the long haul, provided the native people want us there at all. As all recovering alcoholics pray daily, what we need are both courage to change what we can change and patience to accept what we cannot change, but also "the wisdom to know the difference."

I shall address this discussion in three parts, all of them relating to the theme of "burden" in the inculturating Church, and the accompanying qualities of courage and wisdom.[1] Necessarily using a wide brush in this essay, I first turn to the experience of Jesus himself and his disciples in the first generation of the Church; second, I look at the same theme in an overview of mission history; third, I examine the contemporary processes and means for practicing the courage and wisdom needed for inculturation by those working among indigenous peoples.

1 Since, in spite of several fine publications on the meaning of inculturation, there are many differences of opinion on the meaning of the term and even about its value, I will use here an adaptation of the definition given by Pedro Arrupe (See his "Letter to the Whole Society on Inculturation" *Studies in the International Apostolate of Jesuits*, Vol. VII, No.1 (June, 1978), 1–9, and the following articles.) Shortening Arrupe's definition, I see inculturation as a historical prolongation of the mission of the Word in the Prologue of John's gospel: an "incarnation" by which the gospel takes on the clothing of a culture, but also lives and vibrates within that culture in such a way as to lead it further into the "new creation." Readers wishing to probe these meanings further can find excellent brief treatments by J. Peter Schineller, S.J., *A Handbook on Inculturation*, (New York: Paulist Press, 1990), and in the essays in Ary A. Roest Crollius, ed., *Inculturation: Working Papers on Living Faith and Cultures*, Rome: Pontifical Gregorian University, Vol. V (1984).

JESUS AND THE EARLY CHURCH

The gospel accounts show us a dramatic and vibrantly courageous Jesus of Nazareth, who startles the populace of Palestinian Jewish society with his authority and teaching, denounces its distortions of the law and the prophets, and calls it to its wider vocation: the inclusion of the nations in its understanding of salvation, as well as the full inclusion of its own people, its poor, its women, and its marginalized, in its life and culture.[2] Finally (as Luke especially describes) he sets his face toward Jerusalem to die in the consummation of his mission. The Fourth Gospel proclaims to us that all of this story is the story of divine Wisdom coming to a people and dwelling with it so as to show it the glory of the Word abiding in its cultural life—its "tents." One can easily overlook in this dramatic narrative (outside the warmth of the Christmas liturgical texts) the implications of the brief statements of Jesus' own growth in wisdom, stature, and grace, which took place for thirty years in the bosom of his native culture. That is, he too lived through the socializing process that anthropologists call "enculturation." To be sure, his terribly short ministry and his martyrdom snuffed out any further historical development of his "inculturation" within Jewish tradition, at an age at which, I am sure, most of us would not like to be tested for our wisdom! His departure left "the long haul" of proclamation and "inculturation" to those he called to follow him and whom he sent out on mission.

In order to apply a hermeneutic of inculturation to Jesus and his first followers, I would suggest a paraphrase of the first verse of John's gospel: "In the beginning was the mission..." Not to become involved in a tedious debate here on the doctrine of the Trinity and the eternal "mission" of the Son from the Father, one might nonetheless begin with the teaching of the Word sent forth into the world, and indeed, into a specific culture, to dwell in its habitations and share its pilgrimage. Only at its peril does the Church forget that the mission of a Person precedes it into the world. The Church was founded for the mission of God, and

[2] Two missiological works that illustrate this dynamic in the Bible are: Donald Senior, C.P. and Carroll Stuhlmueller, C.P., Biblical *Foundations for Mission*, (Maryknoll, NY: Orbis Books, 1983), and David J. Bosch,, *Transforming Mission: Paradigm Shifts in Theology of Mission*, (Maryknoll, NY: Orbis Books, 1991).

not the mission for the Church. The Word "came into his own," and although his own did not finally accept him, his mission was to show them a greater "glory"—perhaps best illustrated in John 12: 20–24: some "Greeks" ask to see Jesus, and, when he is informed of this, his immediate response is "Now the Son of Man is glorified." Jew and Greek alike have come to meet the Word sent into the world.

It is in relation to this coming of the gentiles to meet Jesus, I suggest, that the "root metaphor" of our entire collection of essays stands out in the ministry of Jesus and the early Church. I refer to Jesus' words, "My yoke (*zygos*) is easy and my burden (*phortion*) is light" (Matt. 11:30). Biblical scholars have commented extensively on this theme: Jesus, while he did not come to destroy but to fulfill the Law, taught it in such a way that it became an easy "yoke" and not the kind of yoke imposed by the current leaders of the Jewish community. Whatever one's systematic theological interpretation of the role and person of Jesus, his temporal mission was to transform the culture of his own birth and heritage into an environment suitable for the Reign of God, to call his people to their full destiny among the nations, in the spirit of the Servant in the book of Isaiah. But such a challenge to any culture, as Jesus well knew, demands both wisdom and courage of the prophet who challenges it. This message was a deep source of chagrin to the leaders of the time, and not without reason.

The pharisees, as a former rabbi colleague of mine constantly reminded me, were not "the bad guys." Still less were they fools. They were the spiritual heirs of the hero family of the Maccabees, who had saved their people from the intolerable yoke of an invading culture. One might easily envision their zeal to support the texts from 2 Maccabees 7, narrating the martyrdom of the mother and her sons for refusal to violate their own cultural principles, which they embraced as principles of religious faith. In that spirit, and no doubt too in the spirit of Ezra and Nehemiah, these leaders were determined to shore up their culture against any further intrusion or dilution or diminution. Tragically, as Jesus constantly demonstrated to them, they did this by imposing their own heavy burden in the form of a rigid application of the Law, and by resisting Jesus' efforts to extend the benefits of the unburdened spirit of the Law to the marginalized and to the wider world. But, as I have said, the pharisees were not fools; they could see the power of ancient cultural symbols and values, however much they may have missed the universalist spirit of those values. In one sense, when the mission Church encountered indigenous leaders of aboriginal cultures who resisted the changes it sought, it was no doubt meeting another version of the Maccabee resistance to foreign intrusion!

The metaphor of the yoke and the burden appears again after the resurrection of Jesus and the coming of the Holy Spirit. It emerges as a problem following upon the conversion and baptism of the household of Cornelius, the pagan, which set in motion a dramatic ingress of pagans into the community. The logical course of the problem involves the imposing of a yoke in the form of a central symbol of Jewish religion and culture—circumcision, and the burden of legal requirements that came with it—which the apostles "and the Holy Spirit" decide is not an absolute but a cultural symbol. Consequently this fledgling Church now enters into a discernment process (we are not told exactly how long it took) to discover what the wisdom of the Spirit might be, and from this issues a courageous proclamation. The Christian leaders decide to lay no "yoke" (*zygon*) upon the gentiles, "which even our forbears could not carry." (Acts 15:10) They prescribe rather only such "weight" (*baros*) as they saw appropriate for all Christians at that moment of dramatic cultural and religious change.

To summarize this situation of the early Church, one may see it as a time of courage, wisdom, discernment, and especially of a search for its own identity and mission. These Christians could see that a courageous decision was demanded of them—to make a startling alteration in cultural religious practice for the sake of the gospel. They needed the courage, not simply because their own synagogue might, and eventually did, excommunicate them, but more deeply because they could no doubt appreciate how profound a shock it is to make such a change in cultural practices. Cultural details may often be a burden, but they also spell security against a hostile world. There is no question about the courage needed by these Jewish Christians to make this decision, or of the wisdom that must govern the decision to undertake the proclamation of a "universalist" message. The mandate now was not simply to accept "proselytes" into the community, as inter-testamental Judaism had been doing, but to alter certain aspects of that community's religious practice.

This pattern of courage and discernment does not simply evanesce in subsequent generations, but rather becomes paradigmatic for the growing Church. One needs only a modicum of church history to appreciate how that community, at first facing martyrdom for its faith, eventually, with the legislation of Constantine and then of Theodosius, acquires first legitimacy and then *establishment* status in the Roman Empire. Examination of this story shows us what Karl Rahner put so succinctly in one of his shorter but more famous essays: namely, that the religious sanctioning of a certain culture surfaces again, and more powerfully, in the European "epoch" of the Church.[3] Inevitably, it seems, the religious community always becomes what Clifford Geertz has called a "cultural system,"[4] or at least a sub-system of a culture. That a Church should bear the marks of such a system is hardly surprising; what becomes problematic and burdensome is that the Community should sacralize that cultural system and simply once again reinstitute cultural demands to be laid as a new yoke on the shoulders of converts.

The Problem in History

Thus, the challenge of wisdom and courage continues in the work of the Church throughout history (a history that has comprised nearly two centuries of the European "epoch"). The dialectic of burden and freedom becomes endemic to the Church as mission, and it *is* a dialectic not to be naively dismissed. Now lest we stoop to that kind of missionary-bashing that comes so easily in the hindsight of current historians and anthropologists, let us acknowledge that, when it comes to courage, we stand on the shoulders of giants. It is a fascinating irony, however, that the courage of the missionaries of earlier generations was powered by factors that most mission scholars no longer see as primary motivation: "no salvation outside the Church"; the absolute necessity of baptism for salvation; the damnation of "pagans." It is even more ironic that this same kind of courage was so often fueled, especially in the nine-

3 See Karl Rahner, "Towards a Fundamental Theological Interpretation of Vatican II," *Theological Studies*, Vol. 40, No. 4 (December, 1979), 716–727.

4 For an understanding of what Geertz means by this term, see his essays: "Religion as a Cultural System" and "Ideology as a Cultural System," in *The Interpretation of Cultures* (New York: Basic Books, 1973), 87–125, 193–233, and "Common Sense as a Cultural System" and "Art as a Cultural System," *Local Knowledge: Further Essays in Interpretive Anthropology*, (New York: Basic Books, 1983), 73–93 and 94–120.

teenth and early twentieth century, by a conviction of the coextensivness of Christianity and "western civilization."

The concern to avoid unbearable burdens and harsh yokes was certainly alive in early mission history. As the Roman Church spread northward and eastward, there was some expansive and moderate spirit in the mission instructions of Pope Gregory I, who warned his missionaries to Britain to avoid churlish smashing of pagan temples and altars, and counseled them to respect local customs. Somewhat later, in the nations of the Slavs, the persistent agitation of Cyril and Methodius obtained concessions from Rome toward the native language in liturgy. But all this being said, we see the establishment of Christianity within European cultures and the converging pattern of Roman theology with Germanic and Gaelic feudalism, Spanish absolutism, and British political and social patterns. This was the gradual birth of "Christendom": the wedding of Christianity to one particular pattern of culture, which gradually and ponderously accumulated to weigh down later missionary effort with new burdens.

Ironically, in this very painful history of imperialism, we find some of the finest examples of discerning wisdom and courage. Amid the atrocities of the *conquista* of New Spain there shines out from the pages of history the example and preaching of those Spaniards who saw the deeper meaning of the gospel—Montecinos, Las Casas, Sahagun and Zumarraga, to mention the best known. In Bartolomé de Las Casas, especially, we see the discerning power to relativize the absoluteness of Aristotelian philosophy as a determinant of Catholic theory and practice, and to use canon law, not as an ultimate norm but as a tool. In the far Asian territories, there are the examples of Matteo Ricci and his colleagues in China, and of Roberto di Nobili in India, all of these quite conscious of the relativity of European culture, and, in the minds of some, even of the superiority of Chinese culture. But again, even in these edifying examples, our hindsight history today can find grounds for criticism of the fact that these Jesuits directed their attention to the "elite" classes of mandarins and brahmins. History is indeed an instrument of prudence, if the historian tries to moderate his or her moral indignation and apply history's lessons to oneself.

Further examples of enlightened policies can be joined to instances of practices that missiologists have come to reject: the prodigious efforts of the Jesuits in South America in the sixteenth to the eighteenth centuries, and those in North America during roughly the same period. Comfortable modern scholarship can find much to criticize in the cultural attitudes of these missionaries, even as they struggled to resist many of the crude aspects of European imperialism against the native peoples. Throughout all of this turbulent history there runs the account of missionaries who did seek to avoid imposing needless burdens and to make the gospel the light yoke that Jesus wanted it to be. We who follow them will certainly alter many of their opinions about what was necessary, but we have much to learn from them about adventuresome courage in encountering foreign cultures.

Our own era of mission and cross-cultural conversation is heir to that massive synthesis of European culture and Christianity that occurred during what has been referred to as "the great century" of mission activity—the nineteenth century, the era of "manifest destiny" in the United States, which colored so dramatically the religious views of missionaries there: the expansionism of white Australia, the identification of the "chosen people" concept with themselves by Afrikaners and other white South Africans, the brutal ideology contained in the concept of *terra nullius* ("no one's land"—a legal term saying that nomads cannot possibly be considered "owners" of land). Our sharing of this history makes us heirs, not to "inculturation" but to a form of "enculturation" in European culture: the Church had become analogous to an infant who unquestioningly grows to full socialization in and domestication within a certain social environment. Without denying the many examples of good services done by missionaries, Catholic and Protestant, to native peoples during this worst period of their history, it is certainly incumbent upon us to study this period carefully. We seek the courage to emulate their generosity, as well as the wisdom to avoid imposing what we now realize are cultural burdens.

The Task of the Contemporary Church

I see the vocation of the Church-as-mission in our epoch as a nurturing of two attitudes: there must be an efficacious penitence for the continued imposition of certain burdens upon local cultures, and there must be an intensification of efforts to help indigenous cultures to experience the gospel as the gentle yoke of Jesus.

Once more, the historical paradigmatic problem of inculturation reflects, I suggest, the essential question encountering the earliest Christians. There is still the problem (and no doubt always will be) of a dominant or conquering culture imposing its political will, and, in some ways, using a religious system as its ally. The analogy breaks down, of course, because in that era of Roman tyranny, the subjugate culture was the one in which and to which Jesus preached deliverance, while for the entire modern period, the religion of Jesus has so often been the partner of the conquerors. It is precisely this conflict that today makes the work of the agents of inculturation, whether they be "outsiders" or (preferably) "locals," so difficult and so much in need of wisdom and courage. Christians involved in such work anywhere must be able to discern where and how to be either "cultural" or "countercultural"—where and how to choose to challenge culture to change or help it to develop along its existing lines.

First, then, how is the Church to be about a counter-cultural process within its own originating culture, in such a way as to remove all needless burdens? Actually, in its official teaching since Vatican Council II, the Church has exercised an impressive capacity for creative thinking, a recognition of the value of distinct cultures and the spiritual validity of religious experience that has not originated in the bosom of the Church as it has known itself previously. The speeches of Pope John Paul II in Canada, Australia, and some nations of Africa have emphasized the image of Christ in his body as Amerindian, Aboriginal or African. At least in the values it *espouses*, the Church is paving the way for the inculturation of the gospel within all dimensions of local cultures, including their religious symbolism. In its *operative* values can it negotiate the delicate and tricky process of respecting particular life-ways and facilitating the passage of local cultures into the life of the "universal" Church?

In this very fledgling period of the "third epoch" the Church cannot continue its earlier policies of "stamping out paganism." Nor can it imitate the example of Jesus preaching to his own local culture that it must lift its needless burdens or open its doors to the world. That is, I, as an outsider, have no credibility when I attempt to preach internal reform to native cultures, or try to exhort them to join in the historical growth of the "mainstream" of society. It would be as if, instead of presenting himself to Peter as a supplicant, the Roman Cornelius had stormed into Joppa preaching that this new community must open itself to Caesar. As it happened, it was a *local*, a member of the conquered culture of no esteem, who offered redemption to the Roman, and who proceeded to reform his own community from within. Where, then, are the leaders, the discerners, the proclaimers speaking for and from the indigenous cultures?

I realize the danger of over-simplifying the problem: a Church that has been European for nearly two millennia must of necessity acquire many characteristics that by now probably cannot be sacrificed without the loss of a symbolism very important to gospel values. Examples might be certain basic credal formulations, the fundamental sacramental symbolism that seems to be represented in the New Testament itself; all of these are issues to be dealt with in shared theological praxis. We can, I suggest, in collaboration between "missionaries" and local leaders, distinguish between the spirituality that vitalizes a community, and the legislation that burdens a community—legislation about government, liturgy, requirements for leadership and ordination that fail to speak to the needs of the communities themselves. Must indigenous peoples turn to the creation of separate "rites" in order to accomplish this, as some have argued? My personal opinion is that such a tedious, lengthy, and expensive process is not necessary. The Roman Rite contains in its original charism the capacity to support diversity—but only if it is prepared to examine itself in its basic charism rather than in its historical accretions.[5]

But let us examine the positive side of this matter. Where in fact is the Church manifesting the courage and wisdom to collaborate with the still marginalized (not necessarily *minorities*, as the present situation of women in the Church demonstrates) in both challenging and supporting them? Surely the most dramatic form of courage can be seen in those Christians who have thrown in their lot with peoples suffering violent repression. These Christians are found in the barrios throughout Latin America, struggling, often against both ecclesiastical and political authorities, to maintain their "basic communities." There are those pastoral workers in the tinder-box townships of South Africa, whose capacities for courage are tested daily by internecine violence and lingering *apartheid*, and whose wisdom is tested by difficult political choices to support one or another group representing the oppressed people. Out of

5 Anscar Chupungco has argued this well in his *Cultural Adaptation of the Liturgy*, (New York: Paulist, 1982).

such situations have emerged, of course, such theological milestone documents as the statements of Medellin and Puebla, the South African *Kairos* document denouncing apartheid, and many other episcopal statements of solidarity with oppressed people.

Turning to our "first world" societies, which have to this point escaped the more violent dimensions of social conflict, we find demands for a quiet courage as well, and for discerning wisdom. Episcopal commissions have gathered data, dialogued with native communities, and made decisions designed to build local churches with respect for local cultural experience, as well as to promote the political aspirations of these groups. Space here allows me to mention only one dramatic case. The last three bishops of the Saulte-Ste. Marie Diocese in northern Ontario, after lengthy consultation with representatives of all communities, have supported a tripartite "sector" system of administration, with Aboriginal, English and French sectors operating both separately and in collaboration.

Church workers who have entered in growing numbers into native ritual life have not needed death-defying courage to do so, but a certain daring is necessary. To participate in native rituals and to encourage experimentation with native symbolism in the liturgy has provoked indignation and resistance on the part of not only some of the hierarchy but of some native Christians themselves, with the native people fearing the possible demonic use of ritual power and the bishops fearing the undercutting of the rubrical support of worship. Even more indignant are traditional native religionists who object to any form of aboriginal symbolism in Christian worship. And yet there is the abiding fact that native ceremonies have the dual symbolism of spiritual healing and inspiration and political support for aboriginal rights.

So what else is new? A survey of the span of history shows us that the same questions abide. We can conclude this brief discussion by again returning to that early community from which all of our history has come: we can turn to the ethical counsels of St. Paul in his letters to the Romans, Corinthians, Ephesians, Colossians, and Philippians. Even in these exhortations there are certain claims that we may see today as merely culturally conditioned advice, such as strict obedience by slaves, or submissiveness of women. But underlying all of these is the spirit that seeks a collaborative and tolerant approach to differences, to avoid the imposition of heavy burdens and unbearable yokes, and ultimately to be governed by the spirit of love. In the discernment and courage necessary to implement its "universal" message, it seems, the Church is *always* "the early Church."

Cross-Cultural Issues *in the* Ministry *of* Spiritual Direction

Susan Rakoczy, IHM

Culture is both social and personal since it is rather like the air we breathe. The air is available to all—the many facets of a particular culture—but each one breathes it to sustain his or her own life. This analogy becomes particularly true when we discuss the cross-cultural dimensions of ministry, specifically spiritual direction. All the principles, caveats, and dynamics operative between two people in this sacred and intimate ministry are influenced by the particular cultural backgrounds of the two persons involved.

As a theologian and director, I have tried to explicate some of the basic principles which undergird the experience of spiritual direction in cross-cultural perspective.[1] In this essay, I wish to make the experience personal by discussing how these principles become enfleshed in the direction experience. I will discuss some aspects of the person as director and three significant issues in cross-cultural spiritual direction: images of God and prayer, discernment, and the interpretation of dreams.

The Person As Director

"Each person is in some respect like all others, like some others, like no other."[2] As a spiritual director, I bring my common humanity as one searching for God, my own American cultural background, now modified by my years in Africa, and the particular strengths and weaknesses of my person to the direction experience.

The gift of Africa came rather unexpectedly into my life in the early 1980's when I was invited to Ghana to teach theology in a diocesan

1 See "Unity, Diversity, and Uniqueness: Foundations of Cross-Cultural Spiritual Direction," in *Common Journey, Different Paths: Spiritual Direction in Cross-Cultural Perspective*, ed. Susan Rakoczy (Maryknoll, New York: Orbis Books, 1992), 9–23.

2 Clyde Kluckhohn and Henry A. Murray, *Personality in Nature, Society and Culture* (New York: Alfred A. Knopf, 1948), 53.

seminary. However, the invitation was withdrawn when a significant number of the seminary staff decided that they did not want a woman teaching theology. This was never said to me directly; rather, I was told that "I was too new to the culture" to teach. But closed doors are often doors to other opportunities, and so I ministered for six years on the staff of the Center for Spiritual Renewal, doing retreats and spiritual direction. I have continued this ministry on a part-time basis for the past ten years in South Africa, in addition to my primary ministry of teaching systematic theology and spirituality.

I came as a self-assured Western woman to Africa in 1982; that self-assurance was broken apart by the impact of life and ministry in a totally new culture.[3] I can never be an African, but I hope I am becoming African by adoption.

Africa has taught me many things as a director. The first is to go slowly and "never take anything for granted" in straining to listen to the ways of God in the heart and life of a person of a different culture. Always we are on sacred ground, and, in Africa, the unity of religion and life which forms the bedrock of African culture is a constant leitmotif in direction. I sense it often in the ways that people speak of the experience of God in ordinary events. In contrast to direction experiences in the United States where I found that people were more inclined to compartmentalize God, prayer, and ordinary experience, in Africa everything and everyone is a potential word of God to a person.

Simple but crucial elements of personal relationship had to be learned. A significant area is non-verbal communication. Sometimes people never look at me during the whole direction session but keep their eyes cast down. This was particularly true in Ghana as a way to show respect to someone in authority or someone older. Initially, it was disconcerting to me, but now I am more comfortable with knowing that lack of eye contact can be shyness or cultural reserve. Another factor linked to this one is the extremely soft and low tone of voice some people use in the direction conversation. Sometimes this shows shyness and ordinary hesitancy, but sometimes it is also a mark of respect. Occasionally I ask people to speak just a little louder when I really cannot understand what they are saying, but generally I place the burden of straining to hear what they are saying on myself.

3 See Susan Rakoczy, "Inner Africa: A Journey of Conversion," *Review for Religious* 50 (1991): 682–696.

Learning to listen in a new way has often meant abandoning time barriers and limits. In a directed retreat experience during my years in Ghana, I learned this lesson very directly. I was accompanying a young Ghanaian sister in an eight-day retreat. Since I was also directing others at the same time, I had allotted about forty-five minutes per person each day for an interview. As the days went by, this sister spoke at greater and greater length: an hour became ninety minutes and then nearly two hours one day. At first I tried to focus the conversation a bit more in order to try to limit the length of the interview (mostly because I had a number of other people to see each day). But that was to no avail. So I relaxed and listened. Towards the end of the retreat the length of our meetings began to diminish to about fifty minutes. At the end of the retreat, when we reviewed the days of prayer, she said to me, "Thank you for listening to me. No one ever listens to me." The next year when she made her retreat, we never met more than about forty minutes; she was satisfied that I could listen to her, and I did.

Since I learned "American English" and its idioms, I have learned to translate various phrases and ideas into "South African English." I find myself translating as I speak since I have the responsibility of making the flow of communication as easy as possible.

As in all spiritual direction relationships, what is most important for the director is to relax, be oneself, and trust the leading of the Spirit. The fact that this person of another culture has asked to see you and has asked you to accompany her or him is already a manifestation of the trust which is at the heart of spiritual direction. It is "too late" to change one's personality, background, strengths, and weaknesses when the person walks through the door. Sensitivity and gentleness are the prerequisites for cross-cultural direction as they are for all spiritual direction.

Images of God and Prayer

In spiritual direction, one treads very lightly indeed when people speak of their images of God. Since God is beyond every image, every verbal construct, whatever words we do use of God are the stuff of mystery, not the product of logical deduction.

Listening to people of different African, European, and Asian cultures over the years, I have heard an amazing variety of images: God is a lion, a breeze, the sun; God is Father, Friend, Mother, Sister,

Grandmother, Listener. God is nothing, gone, darkness, a spark. God takes care of me, will never leave me, is my best friend, is like a school mate. God will listen when no one else will. God is silence, stillness, music, dance. God is the wind in the trees.

When I was in Ghana, I often directed people in weekend retreats. I always begin a retreat with asking people what they hoped for in this retreat: was there something specific on their minds, did they want God to do something particular in this retreat? I was initially disconcerted to hear people say, "I want God to solve all my problems," and then enumerate problems which were overwhelming: an alcoholic spouse, children on drugs, massive financial problems, chronic unemployment. Much for God to do in 48 hours since they expected that these problems would be gone when they returned home on Sunday afternoon!

While I certainly did not want to shake their faith in a provident God who cared for them in every circumstance, I also did not want them to have expectations that could not be met. I learned that if I gently drew them to Scripture texts which spoke of God's care for them in every circumstance, their faith that God was with them even in hardship was strengthened. "God the Supreme Problem Solver" was replaced by "God who will never leave me," including during times that are very hard.

I have spent a good amount of time in direction sessions teaching people how to pray. One would think that perhaps this might be necessary with some lay people who have not had many opportunities for personal formative spiritual growth, but this is also true of priests and religious. I have taught newly ordained priests how to use a simple way to pray the Scriptures, something that their seminary formation totally neglected. One young priest told me that he knew he was supposed to pray the Office, but he had never learned anything more about prayer in eight years of philosophy and theology. He was a bit exceptional, but not unique.

Almost everyone has learned some formal prayers, and persons involved with the charismatic renewal know about communal praise of God in the prayer meeting. But it is often a new thing for women and men, including priests and religious, to learn that they can meet God personally each day in private prayer. The deep spiritual hunger which is so apparent in so many spiritual direction conversations begins to be satisfied with an approach to prayer which starts from their own needs and desires to be with God.

I have also found, somewhat to my delighted surprise, that the wisdom of the great teachers of prayer in our Christian tradition, e.g., Teresa of Avila, Catherine of Siena, John of the Cross, Thomas Merton, speak across cultures and time to Africans in the late twentieth century. I remember a young student commenting that the writings of Merton "speak right to me; it is as if he (Merton) knows the questions in my heart." Merton would be glad of that. It is the same with the other great teachers of prayer: people hear the wisdom and apply it in their own way. One student recently raised the question of how culture-bound Teresa of Avila's writings might be, and, when asked to think about it a bit more, realized that her wisdom, though coming to us in a sixteenth century vessel, was speaking to him.

I have learned as a director to be comfortable with helping people to pray in any way that helps them, as long as they feel they are continuing to search for God. It is a new thing for Africans to be encouraged to dance or sing their prayers, even though music and dance are so central to African culture. Others find various forms of centering prayer helpful, or lying in the grass, close to nature, close to God. In the West, people speak of "creation spirituality" as it is was discovered in the United States in the 1970's. In Africa, the link between God and creation has never been severed, and I have learned much from people in their experiences of finding God in the beauty and awesomeness of nature.

The journey to life, to wholeness, to integration in God and self, which the great teachers of prayer describe, is a universal experience. I sit with people whose lives are darkness in their search for God. God is the smallest ember of what used to be a raging fire of consolation. Women especially are very grateful to find a listening ear and heart for their struggles with self-image, empowerment, and religious maturity. One woman wept when she realized I was not telling her that she was crazy as she shared the dissolution of all her religious images and the disappearance of God. The narration of experiences of sexual abuse by both women and men find their way into the direction conversation. The pain that is revealed is a universal pain, needing compassion and support in the healing process.

In listening to prayer experiences, I have learned to "take nothing for granted" in asking people to describe how they go about prayer. But I have also learned that the boundaries of what can constitute prayer in the search for God have no limits; God is everywhere and always present, including times when the narrative is of loss and darkness.

Discernment

As we are aware, discernment has been discovered or rediscovered in the last generation. While it is true that "everything discerned is nothing discerned," the theological reflection around discernment questions has helped us greatly in understanding what we are trying to do when we ask the questions, "What does God want of me? Where is the Spirit leading us?"

Sometimes discernment language stresses the separation of God and the person so that I/we are searching for the "will of God" outside of ourselves. This framework for discernment is part of the experiences of many people I have accompanied in Africa. Vocation questions are frequently put in the framework of "God's plan for my life." But this is in conflict with the deeper and more pervasive African experience that religion and life are one, that there are no tight boundaries between "ordinary life" and "religious situations." This makes discernment both easier and more complex.

Discernment becomes easier since the person already expects God to be present in their experience and so to "find God's will" happens through reading the clues in their ordinary experience. The role of the ancestors and of various spirits is often very important here.

Discernment is also more complex because of the need to sort out the various strands of experience which might be leading one in a certain direction. It is also more challenging because the Ignatian tradition of discernment, which is a key source of discernment theology and practice, places a great deal of emphasis on affective experience and being aware of one's feelings, either those of consolation or desolation. In some African languages there are no specific words for certain feelings, e.g., joy, anxiety, peace. People do have these feelings; they experience great fear or they feel great happiness. But sometimes they do not have the words to express this in their mother tongue. Since the direction conversation takes place in English (given the number of African languages, the use of an international language such as English, while not the "best" of situations, is the one that is actually possible), the challenge is to help the person link the English terms for feelings with their own experience. One sister told me that during her first directed retreat, her director gave her a sheet with a number of "faces" on it, all showing different emotional states. She found it helpful to look at the faces in order to find a way to express how she was feeling since her mother tongue did not have words for those feelings.

Ancestors and spirits often play crucial roles in discernment. For Africans, an ancestor is a family member who has died and who led an exemplary life while on earth. Not all family members qualify to be ancestors. Depending on the specific ethnic group or tribe, there are different ceremonies, usually held about a year after the death, which confirm a person as an ancestor.

The ancestor's task is to assist the family they have left behind in this life; they are seen as having a mediating function with God, much like canonized saints in Catholicism. The significant difference is that ancestors are linked only to a specific family while the assistance of saints can be sought by all. If tragedy befalls a family, e.g., an accident, a miscarriage, failure in exams, the family examines their attitudes towards the ancestors to ascertain if they have honored them in the prescribed ways. Africans do not worship their ancestors; they venerate them and hold them in very high regard.

It is not unusual for a person to report a dream in which an ancestor told them to do something. A few years ago in the township parish of Mpophomeni where our sisters have ministered, an elderly man in his 80s began to have dreams in which an ancestor told him he should become a Catholic. His whole family was Seventh Day Adventist, so this was indeed a strange message. He shared this with his family, who affirmed the directive. They told him that since the Catholic church helped people in the township and was a "good" church, he should obey the ancestor. He entered the RCIA process and was baptized, confirmed, and receive the Eucharist for the first time one Easter morning.

When people share messages from their ancestors, I try to help them put these experiences in a wider perspective, much as the family did with their grandfather. Sometimes people are ready to do immediately what the ancestor says, and I try to slow down the process, helping them to look at various factors. Sometimes an ancestor comes to them in a dream and warns them or scolds them about something they are doing. A sister described her father coming to her repeatedly, seeming to ask for something. She realized that she had not forgiven him for a way he had hurt her. When she prayed a prayer of forgiveness, the disturbing visits of her father ceased.

Spirits are also forces to be reckoned with in African discernment.

The spirit world of good and evil spirits is generally very real to the African women and men I have accompanied. When I was in Ghana, a priest once prayed in thanksgiving that his faith in Christ had "freed him from the fear of evil spirits." Note that he was not saying that he no longer believed in evil spirits, but rather that he was not afraid of them now. A Western approach to such experiences generally views them as psychological projections of our shadow, of the unhealed parts of our selves.

It is useless to try to explain this kind of experience in Western terms to a person who is genuinely frightened by the sense of evil nearby. The fear limits their ability to pray, makes them anxious, increases fear—holds them in bondage. What is possible is to help them bring their fears to Christ Jesus in prayer and to remind them that the Gospels describe Jesus as having power over evil spirits. I reassure people that the power of Christ is stronger than the power of any evil spirit.

What do I believe is happening to them? I believe that they are having a real experience of evil which threatens their peace and well-being. It does not matter what interpretative framework one uses, the fear is real. And since the power of Christ is real and available to give peace to a troubled spirit, the better strategy is to lead people to pray for the peace of Christ to fill their hearts and console them.

As with visits from ancestors in dreams, the experiences of good and evil spirits must be put in a wider context. The "evil spirit" who is harassing a person may be recognized as my lack of forgiveness towards someone, my bitterness towards a family member. Gentle diagnosis is needed, and extremely gentle analysis of the experience is called for in order to both calm the person and help him or her to deal with their fears.

And "good spirits?" Good spirits are probably angels (and angels seems to be back in favor now). The presence of good spirits helps to increase people's peace and calm. I find that if I accept people's experience of good spirits very matter-of-factly, their importance is not overly magnified. The focus of direction is the presence of God in all situations of life, and if the person speaks of good spirits which help them to deepen their love and commitment to God, all the better.

The pervasive bond between "religion and life" in African experience leads to the expectation that God, who is present in everything and every dimension of life, will be recognized and experienced. It is up to us to be open to divine communication which can come in so many different ways: prayer, people, nature, events, messages from the ancestors, and dreams. African experience of God has strong links with the Ignatian directive of "finding God in all things." The African person already knows that God is in all things.

A special challenge in discernment arises when a person needs to make a decision. Because the spiritual director in Africa is frequently invested with an authority beyond companionship, I have often experienced people saying to me, "Tell me what to do," or "What would you do?" or "What should I do?" In workshops on direction in Ghana, we struggled to help people differentiate between direction as co-discernment, seeking the leading of the Spirit together, and "advice giving."

When such questions are put to me (sometimes they are more subtle, but the intent is there), I try to help the person realize that they have the responsibility for his or herlife. I cannot and will not take it away from them. I assist the person in looking at options, clarifying needs and desires, and urging him or her to take all of this to prayer, seeking the presence of the Spirit in the peace that will confirm the choice. Sometimes this is a real struggle with people who are more dependent, but generally persons accept the responsibility to make their own "decisions in the Lord."

Dreams

We have already begun to speak of dreams as part of the spiritual direction conversation. In my experience as a director in cross-cultural situations, people approach dreams in two ways. People with some familiarity with Western psychology look for symbolic interpretations, whether Freudian, Jungian, or both. Others take the dream at face value and read the dream as a set of instructions telling them what to do. Within the African experience, some dream symbols do have specific meanings, and these act as guides for the interpretation.[4]

[4] In "An African Perspective on Spiritual Direction: An Interview with Sister Michael Mdluli, OP," *Grace and Truth* 12 (April, 1995): 28–31, Sister Michael discusses various dream symbols in Zulu culture. For example, in certain families some dreams indicate specific events. Rushing water can mean death; burning old clothes means that the head of the household has died. Other symbols are not specific to a family: a river to be crossed means

My own framework for working with dreams is basically Jungian, and derives from my own study, from working with my own dreams for a number of years, and through the guidance of my own spiritual director. I am not a clinical psychologist, and I make that clear to persons who ask for more in the relationship than I am qualified to give.

Some directees have had more or less intensive experiences of therapy which have included dream analysis. They freely share their dreams and want assistance in interpreting specific dreams and dream patterns. In all such reflection, the role of the director is to be helpful, not dogmatic. The meaning of a dream is known to the dreamer when an interpretation "clicks" within the person. When the directee senses the meaning of a dream or dream pattern, that information can be very helpful when placed in the larger pattern of their relationship with God.

A man who is husband, father, and priest struggled with a dream in which he was raped in a car accident. As we explored the various themes in the dream and their relationship to the struggles of his life, he suddenly became aware that the driver of the car, to whom he was responsible in ministry and who raped him in the dream, was a symbol of how his marriage and family life were being threatened by the demands of his ministry. As he put it, "The first sacrament—marriage—is being raped by the other sacrament—ministry."

A woman spoke of repeated dreams of climbing a mountain and being threatened by snakes. She tried to link the mountain in the dream with mountains she had known, but things would not connect. Snakes represented fear, and that was clear to her. I tried to expand the dream interpretation a bit and suggested that the mountain might not be a specific mountain, but a particular challenge in her life. That spoke to her strongly since she was struggling with trying to overcome her shyness and hesitancy among people. The snakes suddenly made sense to her not as threats to her physical well-being, but as symbols of her own fears which were indeed harming her sense of herself.

I suggest to people who speak about their dreams rather freely and spontaneously that it is helpful to keep a dream journal in which to note down their dreams in the morning. This can help a person to be aware

---

difficulty in reaching out to others; a locked house shows a problem with relationships.

of very vivid and significant dreams and to note recurrent patterns of dreams.

Sometimes people are surprised when I also suggest that they bring their dreams to prayer and let them be part of their conversation with God, however they are experiencing God. Even for Africans who see God in everything, dreams are more directive to the future than apt material for prayer. But especially when people are disturbed by their dreams, to bring their fears and anxieties to prayer is the very best way of dealing with them.

Conclusion

To accompany a person in their journey in and with God is an enormous privilege. To hear the stories of the working of God in a person's life fills me with awe and wonder, even as very sad stories touch my heart and make it bleed.

My years in Africa, especially as a spiritual director in cross-cultural situations, have given me a double portion of these gifts. I have listened to people from many parts of the world narrate their experience of God and have learned that God's ways are wonderfully diverse and yet have a marvelous unity. I have been constantly amazed at how generously and wholeheartedly people search and find God, experience the nearness of God which increases their hunger and thirst for God, who is nearer to them than their own breath.

Hopefully I have become a better friend of God in hearing the stories of these friends of God.

Ways *of* Living
a Spirituality *for* Ministry

Annice Callahan, RSCJ

How do we experience the light burdens and heavy blessings of ministry? The spirituality of ministry is a difficult topic. By spirituality, I mean our internalized faith-vision, our faith lived in deeds of love and an attitude of hope. By ministry, I mean not only our service to others and God, but also our service to ourselves, self-ministry. What is the relationship between spirituality for personal Christian growth and spirituality for professional involvement in ministry?

People have different spiritualities and theologies. Those we serve represent a pluralism of points of emphasis, interests, and concerns: traditional forms, personal and charismatic approaches, a social peace and justice gospel, an inclusive feminist and creational approach, and an inter-religious "new age" viewpoint. We also need to honor the ethnic diversity of our parishes and communities as we explore how to break bread together. Underpaid immigrants may find it difficult to worship with their wealthy employers. Our ministries are characterized by specific populations, for example, children, adolescents, black people, the chronologically gifted, the developmentally handicapped, Hispanic people, persons living with HIV/AIDS, the poor and homeless, and the separated and divorced.

Theological reflection on our ministry engages the individual believer and the community in a dialogue with scripture and the history of Christian tradition on the one hand, and with the socio-cultural information influencing ministerial concerns on the other hand. Attending to the voice of God wherever it reveals itself leads to a mutual understanding of beliefs and insights. Decision-making then fosters the values and goals of a group's or a parish's mission statement. The art of theological reflection can ground theology in human life experience and puts life experience in dialogue with theology.

We need to keep asking the question: What is the relationship of spirituality, ministry, social justice, ethics, ecofeminism, pastoral care, and church life? In this essay I discuss several ways we may already be living a spirituality for ministry, in order to explore some of the burdens and blessings of ministry. These ways are the following:

- caring for ourselves,
- being real with God and others,
- being ordinary,
- becoming quiet,
- making peace,
- being collaborative,
- becoming people with compassion for global concerns
- becoming one with our universe.

We can reflect on each aspect and learn from one another how to deepen these attitudes of heart in ourselves and in others.

CARING FOR OURSELVES

First, we are called to care for ourselves. Do we? Most of us are probably guilty of abusing our bodies by giving them too much food and drink, and/or too little sleep and relaxation. It seems as if we are hardest on our bodies when we are under the most stress. We need to exercise and take extra sleep when we feel we have the least amount of time for either. What kind of exercise relaxes us? Is it swimming, walking, tennis, racquet ball, or jogging? Hiking or gardening are wonderful ways of communing with the earth in an exchange of energy with it. We feel better about ourselves when we take the time to be good to our bodies.

We also care for ourselves by honoring our dreams, taking the time to record them, to pray over them, and to work with them. Dreams are, I believe, a revelation of God, an unwritten scripture, part of our Judeo-Christian tradition. In Genesis 40–41, Joseph interprets Pharaoh's dreams, and according to the Gospel of Matthew, Joseph has three dreams which he obeys (Matt. 1:18–25; 2:13–23). In fact, our dreams help us to read the book of our own life. We take time to ask God what is being said in the dream, to seek an inspiration, image, or feeling from the dream that might help us on a given day. We try to dialogue in a journal with at least one dream character, asking that character: "Who are you? What part of me are you? What do you want to say to me?" The dialogues may get quite lively!

Besides honoring our night dreams, we need to honor our day dreams, our deep desires. Sometimes we find ourselves having fantasies of how our lives and ourselves might be more effective, more relational, more centered. We can use "active imagination" to get in touch with our conscious dreams, to dialogue with that part of ourselves that wants a voice.[1]

We also care for ourselves by honoring our feelings. That can be difficult. It takes time and trust to get in touch with our feelings. Often it means talking things out with a friend, a counselor, or a spiritual director. For some, fear is a predominant feeling. We need the courage to stroke the very "lion" in front of us that seems so terrifying. For others, anxiety predominates and we need the confidence to trust God at work. For still others, anger is the issue. Sketching, scribbling, or painting can help us express our anger creatively. Feelings have more power over us when left unacknowledged. Accepting and expressing our feelings in appropriate ways is a form of reverence and care.

We need to honor our body rhythms. Some days we have a great deal of energy and ought to focus on difficult projects. Other days we are low on energy. Perhaps that is the time to do work that requires little energy. Honoring our bodies, our dreams, and our feelings are ways of caring for ourselves.

Being Real with God and Others

A second way of living our spirituality for ministry is by "being real" with God and others. Our life of faith is our personal relationship with God. We can exhaust, shame, or belittle ourselves so much that we lose touch with that core faith relationship. We need to take time just to keep in touch with who we really are.

We can go through all the rituals of liturgy and daily prayer without letting God into our daily lives: dreams, feelings, relationships, challenges, and concerns. Keeping in touch with God is an ongoing process of relationship.

We need to share with God not only the peak moments, the joys of life with God, but also our fears, anxieties, and even anger with God.

1 See Robert Johnson, *Inner Work: Using Dreams and Active Imagination for Personal Growth* (San Francisco: Harper & Row, 1989), esp. 135–221.

That is harder to do; it is easier to share our negative feelings with a friend than with God.

We may need to assess our goals relative to our families, parishioners, and friends. Instead of a goal to be always peaceful and loving, our goal might be to be real with God no matter what we are feeling, especially when we are not feeling peaceful and loving. It is harder to "hang in with God" when things are not going our way and people are not behaving as we would like.

Another aspect of being real with God is acknowledging to God how much we long to be accepted unconditionally by one who knows us through and through, how much we long to be held and fed by one in whose presence we feel safe and secure. This was certainly the experience of the psalmist who wrote of being "in the shadow of God's wings," for example, in Psalm 36:7. It was also the experience of the prodigal child who felt held by a forgiving, parental love (Lk. 15: 11–32).

Fostering a real relationship with God may mean adjusting our schedule. Once a month or once a week we might take a day of prayer, getting away from the phone, a visitor, or the mail. We need time to discover our own unique prayer relationship with God instead of reading about how others pray. We need to quiet ourselves by reading, watching trees, listening to music, or taking a long walk alone. We need to learn what quiets us and what helps to nourish that quiet.

It does not make much sense to preach on Sundays or teach on weekdays that people need to discover and trust God's love for them if we ourselves doubt that love. And yet the lifestyle and pace of our lives are not conducive to a simple faith. North American consumerism snares us into buying things we do not need. North American competitiveness drives us to write the best sermons, have the largest Sunday collections and schedule the widest variety of parish activities. All of this takes time and energy, and what gets left behind is the quiet time for nurturing our real relationship with God and with others. Good ministry comes from good ministers who are good to themselves.

Being real with God helps us to be real with others. Listening to others tell their story is a way of valuing experience as the point of departure for exploring God's action in a person's life. In particular, it is important to honor women's experience in scripture, in the history of the church, and in the lives of individual women.[2]

[2] For example, see Margaret Brennan, "Women and Theology: Singing of

We can choose to recall God's deep love. We can focus on that love. All the rest falls into place.

Being Ordinary

A third aspect of a spirituality for ministry is being ordinary. There may be a subtle form of conflict within those of us intent on the spiritual life. A little voice inside of us can nag: "You are supposed to have the most intimate relationship with God. You should have the deepest prayer." Only when we can let go of such unreal expectations and impossibly high standards can we begin to live the spiritual life as it is meant to be lived, that is, to be ordinary. We need to adjust our goals. Instead of searching for the most intimate relationship with God and the deepest prayer, we need to develop a real relationship with God and to settle for "being ordinary." Then our self-esteem is based on a sense of our worth before God, not a sense of our accomplishments in the eyes of others. Ordinary people express their talents and inner motivations without self-consciousness. They invite us to embrace our ordinariness, not merely accept it.[3]

For some, time with God can be boring. Like real life, it is ordinary. Retreat time can be boring, hard as that is to admit. Nothing may happen. In fact, we may be blocked by our feelings; our boredom may be inner anger that has not yet been processed or even acknowledged. We may be out of touch with our dream life. We choose to sit and read books and prayers at God rather than delving into ourselves. We need to let go of our "inner baggage," including the values that our culture promotes: prestige, power, fame, popularity, success, and relevance.

Can we bear to be ordinary? Can we bear to feel the feelings of ordinary people? Can we bear to feel lonely, useless, and empty? Or do we fill every slot in our calendar to feel worthwhile?

Can we bear to be simple, and seek the help we need? Perhaps a perceptive spiritual director who can encourage us to accept and be gentle

---

God in an Alien Land," *The Way Supplement* 53 (1985): 93–103.

3 See Annice Callahan, *Evelyn Underhill: A Spirituality for Daily Living* (Lanham, MD: University Press of America, 1997), esp. 83–101, and Robert J. Wicks, *Touching the Holy: Ordinariness, Self-Esteem, and Friendship* (Notre Dame: Ave Maria, 1994), 13–29, 149–150.

with ourselves is needed: someone to whom we as ministers are accountable, someone who can help us sort out our time commitments and emotional investments. One who struggles with the disparity between a call to be an ordained sacramental minister and functioning in reality as a "plant manager" of a parish, overloaded with administrative detail, would profit from some ministry supervision. We may need a counselor to confront us with the truth, ask challenging questions of us, and help us shift the focus of our lives.[4]

At times we need simply to talk things through with a friend. We can get so caught up with the "hustle and bustle" that we lose touch with the meaning and purpose of our lives. We may still be able to preach and celebrate the sacraments, but we feel hollow, aimless, out of focus. This may be a sign of the beginning of "burnout."

We need to honor the place where we are all ordinary, where we are simply the people of God, where we are just plain human beings trying to live in faith, hope, and love in a troubled world. The underprivileged, who feel very ordinary, might be able, in fact, to help us the most to feel our humanity. "Blessed are the poor in spirit, for theirs is the kingdom of heaven" (Matthew 5:3).

BECOMING QUIET

A fourth aspect of the spirituality of ministry is about becoming quiet. North American culture militates against an atmosphere of quiet. Noise pollution on our streets is but a mirror of the jumble of thoughts and plans in our minds.

Like many North Americans, by reflex we turn on the television, radio, or walkman when we find ourselves alone. Do we drive the car without the radio or a tape playing? Do we get to church early in order

[4] On priests who benefit from spiritual direction, see Susan Howatch, *Glittering Images* (Glasgow: Collins, 1987); and *Glamorous Powers* (NY: Fawcett Crest, 1988).

On the question of a minister's accountability, see Urban T. Holmes III, *Spirituality for Ministry* (San Francisco: Harper & Row, 1982), 187.

On creative images of the priest, see, for example, Urban T. Holmes III, *Ministry and Imagination* (New York: Seabury, a Crossroad Book, 1976), 219–242.

to become centered before worshipping or leading others in worship? Do we take time during the day to pause and be still?

Today we speak of becoming quiet as a feminine image: the image of pregnancy and birth, the image of waiting. We need to wait to let things happen, to allow babies come to birth, to allow relationships to come to birth, to allow ideas to come to birth. We need to be quiet and listen.

To become quiet is difficult when we feel afraid, anxious, or angry. Yet that is the moment we most need to be quiet. We can learn to let our irritations be the "pearls of great price." An oyster coats an irritating piece of sand with mucous and it becomes a pearl. When irritations come, we can lean back on our experience of peace and healing and act from our center of peace.

This attitude of becoming quiet colors our choice of leisure activities. Do we choose to relax in the middle of a noisy crowd? Or to relax at a quiet park?

Relaxing our minds reduces stress. Deep breathing can help; some make a habit of pausing at their desk to breathe deeply for several minutes during their day. For others, a half-hour prayerful walk after lunch can help them gain new energy and hope. For still others, centering prayer is what quiets them.[5] A helpful practice to nurture quiet is what used to be called "spiritual reading," a half-hour a day of reading. For those of us in ministry in the area of theology, it may be helpful to turn to novels, self-help books, or movies to supplement our reading. From that place of quiet deep inside of us, others can draw nourishment and comfort: "In quietness and confidence shall be your strength" (Is. 30:15).

5 On deep breathing, see Lucy Lidell, *The Sivananda Companion to Yoga* (New York: Simon & Schuster, a Fireside Book, 1983), 69–77; and Nancy Roth, *The Breath of God: An Approach to Prayer* (Cambridge, Ma.: Cowley, 1990).

On reducing stress, see Robert J. Wicks, *Handbook of Spirituality for Ministers* (New York: Crossroad, 1995), 249–258.

On centering prayer, see Thomas Keating, *Intimacy with God* (New York: Crossroad, 1994); *Invitation to Love* (Rockport, Mass.: Element, 1992), and *Open Mind, Open Heart: The Contemplative Dimension of the Gospel* (Amity, NY: Amity House, 1986).

MAKING PEACE

A spirituality of ministry today is based on genuine gestures of peacemaking, such as apology, forgiveness, and reconciliation. Often this begins with forgiving ourselves and letting go of false guilt. Sometimes we need to let our feet be washed before we can wash others' feet.

We all suffer from childhood pain, family pain, or institutional pain. We are all in need of healing and redemption. At different points in our lives of ministry in the church, we need to turn to God and beg God to be parent, lover, or best friend. We need to celebrate ourselves and our God, the God of surprises. We need to know our limits and not expect to surpass them.

The needs of others press in upon us. If we do not take care of ourselves and hold up our own pain for healing, we may not be able to care for others in their pain. Part of our pain may be the aging process as we move through midlife toward becoming senior citizens. Accepting the diminishment of our parents, siblings, and friends is also part of our pain. The shame that addictions deny, the shame of having been abandoned, the shame of emotionally abandoning ourselves may be our pain. By allowing ourselves to feel our pain and let go of it, we become "wounded healers" for one another.[6]

An important part of making peace is a commitment to nonviolence. Peace is the sign of the disarmed heart. A commitment to nonviolence promotes equality, honesty, peace, and justice for all. It resists classism, militarism, racism, and sexism in ways that are not passive but rather nondestructive. This commitment to nonviolence has individual, interpersonal, local, and global implications for promoting peace and justice.[7]

6 See, for example, Mary D'Apice, *Noon to Nightfall: A Journey through Midlife and Aging* (Melbourne: Collins Dove, 1989); and Dorothy Sennett, ed., *Vital Signs: International Stories on Aging* (Saint Paul, Minnesota: Graywolf Press, 1991). See also Henri Nouwen, *The Wounded Healer* (Garden City: Doubleday, 1979); and H. Nouwen, D. McNeill, D. Morrison, *Compassion: A Reflection on the Christian Life* (Garden City, NY: Doubleday Image, 1983). Cf. Annice Callahan, *Spiritualities of the Heart* (New York: Paulist, 1990), 201–217, and *Spiritual Guides for Today* (New York: Crossroad, 1992), 117–135, 166–169.

7 See Joan Chittister, *Wisdom Distilled from the Daily: Living the Rule of St. Benedict Today* (San Francisco: Harper & Row, 1990) 186, 188–190.

Those around us draw strength from our peace. We discover the fulfillment of the promise: "I will extend peace to her like a river" (Is. 66:12).

Being Collaborative

Ministry viewed as partnership in Christ means being collaborative. Returning leadership to the Christian community means accepting our limits in ministry, recognizing leadership as an activity of the group, fostering collaboration, and empowering laypeople to exercise their stewardship. Purifying church structures involves fostering the powers of the weak and sharing the powers of the strong by developing processes that expand dialogue, extend participation in decision-making, and enlarge mutual accountability.[8]

Contemporary images of interdependence and mutuality shed light on a new vision of Christian leadership and challenge a vertical view of leadership. A network of plural strengths reinforcing our need for one another's resources stands in marked contrast with a hierarchy of strong and weak. Historically, service has become authority when the function of serving has been made into a social status and community ministries have been hierarchized. The emphasis on ministry as partnership in Christ recasts the role of the religious leader as one of fostering teamwork and tapping of each participant's resources rather than fostering dependence and draining the leader's resources. The leader's role is to help the group act effectively by nurturing commitment and fostering collaboration. Effective leadership is about empowerment.[9]

Prophetic leadership in an adult church means exercising ministry as partners not as parents. Partnership implies acknowledging mutual needs, sharing power, and managing conflict. Healing the wounds of misused authority includes exploring and implementing the symbolic role of the leader to serve and mediate the presence of God, not provid-

8 See Evelyn Eaton and James Whitehead, *The Emerging Laity* (Garden City, NY: Doubleday, 1986), esp. Chapters Six, Ten, and Eleven; and *Community of Faith* (Minneapolis: Winston-Seabury, 1982), esp. Chapters Five and Six. Cf. J. Gordon Myers, "Decision Making—Goal of Reflection in Ministry," in *Method in Ministry*, eds. James D. Whitehead & Evelyn Eaton Whitehead (Kansas City, Mo.: Sheed & Ward, 1995), 99–112.

9 See Evelyn Eaton and James Whitehead, *The Emerging Laity*, 76–102.

ing it. According to the principle of sacramentality, the minister is a sign of God's grace, not only an instrument of God's grace. Ministerial spirituality holds that a minister is in touch with the reality signified and lives by the values represented and proclaimed.[10]

Pastoral ministries contribute to leadership within the local church community. Questions about the nature and function of ministerial priesthood impinge upon questions about the nature and function of the deaconate, as well as the nature and function of pastoral assistants in a parish. Does sacramental ordination alone constitute a community leader? What is the sacramental nature and function of the pastoral assistant as community leader in a priestless community? How is the sacramental ordination of women priests, women bishops, and married people strengthening Anglican and Protestant parish communities? How are married Roman Catholic deacons contributing to parish life? What are the issues preventing the priestly ordination of Roman Catholic women and of married Roman Catholics? How does the postmodern insight that we create reality affect church structures?

Pastoral teams provide integrated leadership whereby non-ordained and ordained ministers can effectively build up the local church community, recognizing and respecting the implications of cultural and ethnic pluralism, as well as of ecumenical and interfaith dialogue. An effective pastoral team provides spiritual fellowship and an opportunity for theological reflection. The necessity for collaboration among churchpeople in the life and authority of the church is awakening the laity to their identity as the church and is provoking structural changes regarding the nature and function of priests, pastoral assistants, parishes, bishops, dioceses, the pope, and the college of bishops. The church of the future must grow from the grass roots of base Christian communities; it must be ecumenical in its objectives, democratic in its decision-making, and sociopolitical in its commitments. Such a church makes God's forgiving nearness visible in an un-Christian world.[11]

10 See, for example, Evelyn Eaton and James Whitehead, *The Promise of Partnership* (HarperSanFrancisco, 1991), esp. Parts One, Three, and Four; and Richard P. McBrien, *Ministry* (San Francisco: Harper & Row, 1987), esp. Chapter Four on ministerial spirituality.

11 See Karl Rahner, *The Shape of the Church to Come*, trans. Edward Quinn (New York: Crossroad, 1983), 93–132; and Edward Schillebeeckx, *Ministry* (New York: Crossroad, 1981), 134–142. Cf. Brian O. McDermott, "The

One form of ministry to others as partnership in Christ is spiritual companionship found in groups or in soul friendship. Such partnership ministry is also found in basic Christian communities which gather to read and reflect on the gospel and give us a model for priestless parishes.

Another form of ministry in partnership is hospitality evangelism which replaces evangelistic meetings. A fifth form is the ministry of groups of theologically trained pastoral assistants. A sixth form is the group trained in lay ministry to the aging, the sick, or youth, often modeled on Clinical Pastoral Education.[12]

Two forms of partnership ministry which are part of a Christian feminist vision of ministry are friendship and solidarity. In friendship we find comfort in suffering, courage to take risks, and joy in another's presence. Solidarity with the poor silently works to change structures of inequality and to transform systems of oppression.[13]

Still another other way to embody ministry as partnership in Christ is to transform church boards into communities of spiritual leaders. And yet another way is to encourage hands-on leadership in community ministry.[14]

Becoming Poeple with Compassion for Global Concerns

We are called to become people with compassion for global concerns. In collaboration with others, we commit ourselves to pray for others, to let our hearts reach out in compassionate solidarity with poor and oppressed people.

Relationship among Authority, Leadership, and Spirituality in Ministry," in *Handbook of Spirituality for Ministers*, ed. Robert J. Wicks, 381–390.

12 See Regis A. Duffy, *A Roman Catholic Theology of Pastoral Care* (Philadelphia: Fortress, 1983), 85–115; Donald Peel, *The Ministry of Listening* (Toronto: Anglican Book Centre, 1980); and James N. Poling and Donald E. Miller, *Foundations for a Practical Theology of Ministry* (Nashville: Abingdon Press, 1985), 126–146.

13 See Lynn N. Rhodes, *Co-Creating: A Feminist Vision of Ministry* (Philadelphia: Westminster Press, 1987), 122–126.

14 See Charles M. Olsen, *Transforming Church Boards into Communities of Spiritual Leaders* (Bethesda, Md.: Alban, 1996); and Carl S. Dudley, *Next Steps in Community Ministry: Hands-On Leadership* (Bethesda, Md.: Alban, 1996).

Compassion includes the genuine acceptance and affirmation of others, knowing that their strengths and weaknesses are somehow are own. Compassion also includes the ability to respond to another in pain when we ourselves are in pain, like Jesus who took pity on the crowds and spoke with them even though he was still feeling keenly the death of John the Baptist (Matt. 14:13–14). When we reach out to respond to another's need even when we are feeling needy, we are given a unique power to love, a rare compassion that sensitizes us quickly to the other and offers encouragement.

The compassion we show others springs from the heart of God. The root word of compassion is *rahamim*, "womb-love," from the root word *rehem* meaning womb. God feels for us with "motherly-compassion."[15] This is how deeply we can feel for others, how deeply we can suffer with them. This is also how deeply we can feel for the earth, how deeply we can suffer with it. Compassion implies healing the environment—individual, social, ecological, political, or cosmic: "When you work in cooperation with others motivated by compassion and using thrifty, artistic means, your actions send ripples of positive change in every direction."[16] Through our compassion we mediate God's forgiveness for another. Mercy is the feminine face of God, the feminine wisdom of God, sophia. This sophia is "the tenderness with which the infinitely mysterious power of pardon turns the darkness of our sins into the power of grace."[17] Mercy identifies with God's compassion which is "a transforming energy of social relatedness."[18] We can liberate this divine energy by calling on our resources for social compassion in the power of connectedness, friendship, and the social bonding of feeding and being fed, all symbolized and mediated by the Eucharist.

15 See Phyllis Trible, *God and the Rhetoric of Sexuality* (Philadelphia: Fortress Press, 1978), 31–59, esp. 45.

16 Eknath Easwaren, *The Compassionate Universe* (Petaluma, Ca.: Nilgiri Press, 1989), 128.

17 Thomas Merton, *The Complete Poems of Thomas Merton* (New York: New Direction, 1977), 366.
On the mystery of God as sophia, see, for example, Elizabeth A. Johnson, *She Who Is* (New York: Crossroad, 1992), 124–187, 293–298. Cf. Dennis Edwards, *Jesus the Wisdom of God* (Maryknoll, NY: Orbis, 1995).

18 See Rosemary Haughton, "Liberating the Divine Energy," *Living with Apocalypse*, ed. Tilden H. Edwards (San Francisco: Harper & Row, 1984), 75–89.

Compassion goes beyond pity as we move into solidarity. Compassion as involvement in the situation becomes remedial activity against oppression and suffering beyond the passivity of discouragement. Solidarity is the basis for an authentic ecclesial identity of the church of the people, not merely a protectionist church for the people.

Viewed as a nonviolent form of criticism, compassion grieves with the grieving and helps free them to be energized to love; a violent criticism can discourage and cause dissension. "Compassion constitutes a radical form of criticism, for it announces that the hurt is to be taken seriously, that the hurt is not to be accepted as normal and natural but it is an abnormal and unacceptable condition for humanness."[19]

Solidarity with the oppressed implies liberating the oppressors from their ambition, power, and selfishness. To opt to combat the oppressive class is to side with the oppressed and love our enemies. In fact, "our love is not authentic if it does not take the path of class solidarity and social struggle."[20]

Today we are beset with global concerns. We need to let our hearts be stretched to feel one with others in our global community who are poor and oppressed, abused and marginalized. We need to remember that God brought us out of the house of slavery into the promised land, out of our addictions and compulsions into the promised land of personal relationship with a compassionate God (Exodus 5:6–9; Deut. 4:37–38, 7:18–19). We need to remind others that God has brought them into this promised land of relationship. We need to remind ourselves that we are called to be people of compassion. We can only live our mission of compassion if we are gentle with our selves, others, God, and the universe in a quiet way.

Becoming One with Our Universe

An attitude of gentle reverence frees us to become one with our universe. Believing the new story of creation, we can view God as pre-

19 Walter Brueggemann, *The Prophetic Imagination* (Philadelphia: Fortress, 1981), 85.

20 Gustavo Gutierrez, *A Theology of Liberation*, trans. Sr. Caridad Inda and John Eagleson (Maryknoll, NY: Orbis, 1973), 275. See also Jon Sobrino, *The Principle of Mercy* (Maryknoll, NY: Orbis, 1994), 15, 144–160.

sent in the evolutionary process of the universe, including those of our planet, in such a way that we regard the universe as God's body, the visible sacrament of the invisible God, and ourselves as God's partners.[21]

We can choose to enhance human beings and the earth's resources, not exploit them. We can be willing to take responsibility for the future of our children. By so doing we prioritize the quality of human life rather than the quantity of goods. We become growingly sensitive to integration, interconnectedness, and interdependence. We encourage bodily expressions in rituals. We reverence all, including the sacredness of sexuality. We recognize that the neglect of the feminine principle is the root of alienation from the earth.[22]

We can perceive the need for a holistic vision of human society based on the integration of spirituality and politics. This vision explores alternative energy sources to cut down on pollution, different methods of food production and distribution to avoid depleting the land, and the recycling of waste to eliminate forms of disposable waste.[23]

The cosmos is a network of interdependent beings collaborating together in their future. God's plan for the new creation is universal solidarity not self-interest. Universal solidarity implies the solidarity of all people with the non-human community, with air, water, and the earth itself, loving our neighbors as part of ourselves in intertwined destinies.[24] We become convinced of interdependence and commit ourselves to live

[21] See, for example, Brian Swimme and Thomas Berry, *The Universe Story* (San Francisco: HarperSanFrancisco, 1991). Cf. Brian Swimme, "The Cosmic Creation Story," *Readings in Ecology and Feminist Theology*, ed. Mary Heather MacKinnon and Moni McIntyre (Kansas City: Sheed & Ward, 1995), 249–258. See also Sallie McFague, "An Earthly Theological Agenda," in *Ecofeminism and the Sacred*, ed. Carol J. Adams (New York: Continuum, 1993), 84–98. Cf. Sallie McFague, *The Body of God* (Minneapolis: Augsburg Fortress Press, 1993).

[22] On this point, see, for example, Margaret Brennan, "Patriarchy: The Root of Alienation from the Earth?" *Thomas Berry and the New Cosmology*, eds. Anne Lonergan and Caroline Richards (Mystic, Ct.: Twenty–Third, 1987), 57–63.

[23] See, for example, Rosemary Radford Ruether, *Gaia & God* (San Francisco, 1992), 254–74, 305–6.

[24] See Jay B. McDaniel, *Earth, Sky, Gods and Mortals* (Mystic: Twenty-Third Pub., 1990), 29.

reverently in the new creation, feeling a kinship with the earth and with the universe.

Conclusion

A North American spirituality for ministry must be rooted in our personal, communal, and ecological relationship with God. Caring for ourselves, serving others, and feeling our kinship with the earth are ways for us to become people of contemplation and communion at one with our universe. Becoming real, ordinary, quiet, and compassionate are counter-cultural attitudes in our consumer, technological society, a society which feeds our false personae and competitive struggle for success. Our point of departure for helping others is our inner wisdom born of reflection in faith on our experience in dialogue with the paschal mystery of Christ being lived out in the signs of the times. Contemplating others and our ravaged earth with the compassionate gaze of Jesus leads us to embody and educate to reconciliation, justice, and peace at interpersonal, local, and global levels. Ministry is a light burden with heavy blessings once we allow ourselves to be both liberated and liberating.

Towards Doing Theology *in a* Parallel Culture

Mary Jo Leddy

The following reflections are exploratory in nature. My concern is at once theoretical, pastoral and political. If we make the judgment that our present economic system is soul-destroying, destructive of community and what is best in our tradition, then what are we to do? Can we or should we try to change the system? Or should we seek refuge from it? Should we focus on trying to create spiritual and economic alternatives to it?

In other words, what is the possible stance of Christians in regard to this culture? What could or should be the relationship of the church to this particular culture? These are questions which Margaret Brennan has asked with courage and integrity in her vocation as a theological educator.

In exploring these questions, I will begin by describing the dispiriting dynamic of the culture of money. Then, I want to review H. Richard Niebuhr's now classic formulation of the five types of relationship between Christ and Culture.

It seems important to examine Niebuhr's preferred model, that of Christ as the Transformer of Culture, in order to reflect on its adequacy as a theological position in relationship to the culture shaped by the neo-liberal economy. Let me indicate at the outset that I tend to agree with Stanley Hauerwas' critique of Niebuhr and his contention that Niebuhr dismissed the model of Christ Against Culture too quickly. Yet I find Hauerwas' position, that the Church could or should be an "alternative culture," somewhat unsatisfying.

In my view, the reflections of the Czech writer and political leader Vaclav Havel, on the possibility and necessity of a "parallel culture," are far more grounded and hold greater potential for a real pastoral, theological and political response to the culture shaped by a neo-liberal economy.

A Dispiriting Economy

The culture we live in is profoundly shaped by economics. It is the culture of money. It is materialistic, soul-destroying, dispiriting. I suspect that most theologically minded people would be conscious of how the message of the gospel is dissipated by the more conspicuous forms of consumerism. However, I think that we need to articulate the deep destructiveness of the present economic system, its capacity to diminish and distort the spirit of a human being. Such an articulation is necessary, I believe, because spirituality is being used by some as a refuge from the political and economic uncertainties of today.

Let me briefly evoke what I see to be the soul-destroying dynamic of late industrial capitalism: Capitalism is driven by consumption. Without consumption, there is no production. We become complicit consumers to the extent that we listen to the words and see the images propagated through advertising. We get the message that we must have *more* to be more. We must have *more* to be happy.

Advertising sets up within us a craving for more—more things, more relationships, more experiences, even more spirituality. What once seemed like a luxury becomes a need, and then even a basic need which seems almost "second nature." The human spirit becomes consumed by such craving, to the extent that it is not uncommon to hear a young person say "I will die if I don't have it."

We get the message that it is *never enough*. Slowly but surely this message transmutes and transforms us at other levels of our consciousness: "I don't have enough" becomes "I am not enough" and becomes "I am not good enough."

- "I am not enough." A generalized sense of powerlessness.
- "I am not good enough." A vague feeling of guilt.

I would suggest that this intrinsic link between the dynamics of the economy, the psyche and the spirit goes some way to explaining the perplexing situation that many people in North American culture feel generally powerless and vaguely guilty. And this in the richest and most powerful culture on earth. Note that this general sense of powerlessness is not necessarily attached to any specific experience, and the vague feelings of guilt are not akin to feelings of real responsibility.

The culture of money has its own psychological and spiritual ideologies which serve to keep the present economic system intact. I think that we as church people tend to treat powerlessness, vague guilt and personal unhappiness as personal and psychological problems. Sometimes we see spirituality as a solution to those problems. If our pastoral work accepts the present economic system as "reality," then the fundamental lie at the basic of the culture of money goes unquestioned.

The present economic system is based on the myth, the lie, that if we have more we will be happy. It is a myth which seems to get stronger as the west begins to decline ever more. Our culture has been profoundly shaped by the myths of modernity, the myths of progress and technology. For the so called "ordinary person" these myths are sometimes articulated in everyday beliefs such as: "If we just work hard enough and smart enough, then things will be better and better tomorrow."

This secular confidence is now being shattered. (I think it was irrevocably shattered after Auschwitz and Hiroshima. However, it is now being experienced "in the streets" as it were.) As the false hope for a better and *better* world fades, it is replaced by the illusion that if we just have *more* and *more* we will be happy. The secular version of meaning and purpose has become thoroughly materialized in the craving for more and more.

This craving, of course, cannot be satisfied. The consumer culture works only as long as we remain dissatisfied with what we have or are. It is never enough. To the extent that we allow the culture of money to define happiness for us, we will remain unhappy, perpetually dissatisfied. When this intrinsic dissatisfaction is coupled with the an economic crisis, can a spiritual crisis be far behind? If I must have more to be happy, what are we to do in a time of "less," of downsizing and economic diminishment?

The cravings induced by the culture of consumerism eat away at the most fundamental religious attitude, the attitude of gratitude—without which prayer and worship and generous action are impossible. When we dwell in a sense of gratitude we can say "it is enough" and we can become more sensitive to those who really do not have enough, not enough of the daily bread which prompts spontaneous gratitude.

The economy of grace is radically different from "the economy." In the economy of grace, God's love is given to all, forever and for free. In the culture of money, we are educated to adjust to what is called "reality" and in this reality "you can't get something for nothing."

However, as we grow into faith in the economy of grace, we begin to understand that God loves us for nothing, for no reason—not for what we have, for what we do, or for how we look, but just because we are. And because God loves us, we are. The mature religious response to this experience of the gratuity of God's love is, as Gustavo Guttierez has said, to begin to want to love God for nothing.

It is difficult to be grateful in an economy driven by greed. It is difficult to be happy in an economy which perpetuates itself only through culturally induced dissatisfaction.

Elsewhere I have written about the relationship between the economics of capitalism and the political system of liberal democracy.[1] I have attempted to describe how the uncritical acceptance of the socio-political forms of liberalism has become problematic for religious and church communities. Liberals believe that, as in economics, so too in various socio-political dynamics, the "invisible hand" of the dynamics and process of interaction will produce the greatest good for the greatest number of people. Salvation and meaning will arise from interaction. Community will arise from the interaction of individuals. However, in the experience of liberal cultures the most we can hope for is the agreement to disagree politely. I think we are beginning to see that liberal political and social illusions can be sustained only as long as the economic pie continues to expand. As long as the pie gets bigger, we can hope that there will be more for everyone. However, when the pie becomes more limited or even begins to shrink, then conflicts arise between various interest groups and individuals. The invisible hand, in whatever glove, has no commonly agreed upon basis for deciding between such conflicting rights and interests. A notion such as the "common good" was neither the foundation nor the possible goal of liberal socio-economic theory.

These very brief considerations lead one to conclude that it is very difficult to be "in" this culture of money and its corresponding political system without almost becoming "of" it. What then could or should be the relationship between Christ and Culture, between the church and culture?

1 cf. Mary Jo Leddy, *Reweaving Religious Life: Beyond the Liberal Model* (Mystic, Conn.: Twenty-Third Publications, 1985).

A Review of Niebuhr's Reflections on Christ and Culture

For H. Richard Niebuhr, the relationship between Christ and Culture was "The Enduring Problem." He notes in his introductory remarks that, "the repeated struggles of Christians with this problem have yielded no single Christian answer, but only a series of typical answers which together, for faith, represent phases of the strategy of the militant church in the world."[2] Based on Ernst Troeltsch's categories of sect versus church, Niebuhr created a typology of five "typical answers."

Christ Against Culture or the opposition between the church and culture: This model has arisen when people of faith believe that they are faced with an either-or decision. In the early Christian period this was found in the antagonism of Christians to Jewish culture or in flight from the Graeco-Roman Empire. In the medieval period, monastic orders and sectarian movements called on believers to separate themselves from the world. Niebuhr characterizes these groups as "little groups of withdrawing Christians."[3]

For reasons which may be a little clearer further on, I want to examine Niebuhr's discussion of this particular type at greater length than his exposition of the other models. He views the near total rejection of culture in someone such as Tertullian as coinciding with the notion that original sin is transmitted through society. Niebuhr claims that such a position characterized the monastic movement and says that St. Benedict was in the tradition of an exclusive Christianity. Whatever contribution the monks made to culture was incidental and unintended—according to Niebuhr. He also discusses Protestant sectarianism, specifically the Mennonites and Quakers.

Niebuhr recognizes some good in this type. Christian withdrawals from the world have often been of great importance both to the church and to the culture. They have often led to reformations in the church and in the world—although this was not their primary intention.

Nevertheless, for Niebuhr, this withdrawal from and renunciation of the world must be followed by responsible engagement in the world if it

2 H. Richard Niebuhr, *Christ and Culture.* (New York: Harper & Row, 1951), 2.

3 Ibid. 41.

is not to degenerate into a selfish quest for personal perfection and peace. Niebuhr goes on to say that it is impossible for a person to live outside of a culture. A person cannot even think or speak outside of culture because culture is inside the person. One can reject culture only by using categories of thought derived from that culture.

The rejection of culture takes place when the culture is seen as corrupt and uncultivated nature is seen as less corrupt. One of the dangers in such a position, according to Niebuhr, is that sin is in the world but not in the Christian. The temptation of a Manichean dualism haunts this model: a corrupt world divided from the world and a spiritual realm governed by Christ. There is a tendency to spiritualism, to a loss of contact with the historical Jesus, an abandonment of the scriptures and the scriptural Christ—a disconnection with the body of Christ.

Christ and Culture or the recognition of a fundamental agreement between Christ and Culture: In this model, Christ appears as the great cultural hero embodying the best of a culture. Niebuhr gives examples of historical responses characteristic of this type:

> "Answers of this kind are given by Christians who note the close relation between Christianity and Western civilization, between Jesus' teachings and democratic institutions; yet there are occasional interpretations that emphasize the agreement between Christ and Eastern culture as well as some that tend to identify him with the spirit of Marxian society."[4]

In his first and second models, Niebuhr has presented what he sees as the two extremes of the relationship between Christ and Culture.

Christ Above Culture. Niebuhr sees Thomas Aquinas as representative of this type of answer to the enduring problem. (I suspect there are Thomists who would seriously object to this) Although culture can lead people to Christ in a preliminary fashion, it is only when Christ enters into culture from above "with gifts which human aspiration has not envisioned and which human effort cannot attain"[5] that human beings are related to a supernatural society and a new value center.

Christ and Culture in Paradox: In this type, the human being is seen as subject to two moralities, a citizen of two worlds which are not only discontinuous, but also opposed. "In the polarity and tension of

4 Ibid.

5 Ibid. 42.

Christ and culture life must be lived precariously and sinfully in the hope of a justification which lies beyond history."[6] For, Niebuhr, Luther is the greatest representative of this type.

Christ the Transformer of Culture: This is clearly Niebuhr's preferred type. The opposition between Christ and the world is recognized but the antithesis does not lead either to Christian separation from the world or to mere endurance in the expectation of a transhistorical salvation. Christ is seen as the converter of society. The key representatives of the model are Augustine and Calvin.

It is, of course, always risky to ask what strategy Niebuhr would suggest to the church in a culture shaped by neo-liberal politics and economics. He might say that human beings can and should change such systems. He might also point out that such a culture is not totally corrupt and that there still exists within it points of possible transformation. However, we do not know what position Niebuhr himself would advocate today. The question we can ask is whether the realities of the culture of money in the late twentieth century call for a "strategy" which Niebuhr did not or could not envision?

A Critique of Niebuhr

In recent years, a stinging critique of Niebuhr's formulation of the "enduring problem" has been issued by the theologian Stanley Hauerwas. This critique has been widely discussed in the lively conferences of The Gospel and Our Culture Network.[7]

Hauerwas holds that the "Christ Transforming Culture" model of church is (in spite of Niebuhr's qualification that liberals belonged in the Christ of Culture model) that of liberal mainline American Protestantism which sought to transform America into something better. It is Hauerwas' position that "few books have been a greater hindrance to an accurate assessment of our situation than *Christ and Culture*."[8]

This critique is based on Hauerwas' judgment that we are living in a post-Christendom era which demands a reassessment of the church's

6 Ibid., 43.

7 cf. George R. Hunsberger and Craig Van Gelder (eds.), *Between Gospel and Culture*. Grand Rapids: Wm. B.Eerdmans Publishing Co., 1996.

8 Stanley Hauerwas, *Resident Aliens*. Nashville: Abingdon Press, 1989. 40.

relationship to culture. Niebuhr's confidence in the creating and redeeming activity of God "had the effect of endorsing a Constantinian social strategy."

According to Hauerwas, in Niebuhr's strategy the church is a confident world-affirming church which believes that people can transform the world through power and choice into something better. This church was opposed to the world-denying sect described by Niebuhr in his first model. Hauwerwas rejects the all-or-nothing choice between choosing to be with all of the culture or becoming a sectarian nothing. For Hauwerwas, the problem with this approach is that, in the process of trying to change the world, the church itself has become transformed. This Constantinian approach was shared by both conservative and liberal churches, by the left and the right, who felt that they had to get involved in politics—to influence those with political power so that they could transform the world.

Hauerwas' position is that the church must become a polis or an alternative culture. Instead of seeking to transform the world, the church must first become the church, must become converted to the gospel. This is, for him, the exciting possibility opened in a post-Christendom period. He clearly is seeking to reclaim some of the positive dimensions of "the sect" which Niebuhr had so criticized. Hauwerwas' proclamation of the particularities of a gospel community, of narrative ethics, is well-known and much discussed. (Within the limits of this paper I cannot discuss all of the aspects of Hauerwas' position, so I will restrict myself to his specific critique of Niebuhr.)

It is important to note that Hauwewas is not advocating that Christians flee the world, because for him, there is simply no place else to be. "The church's only concern is how to be in the world, in what form, for what purpose."[9] Hauwerwas seeks to reclaim the church as a confessing church—in continuing solidarity with the confessing church of Germany at the time of Hitler. "(T)he confessing church finds its main political task to lie, not in the personal transformation of individual hearts or the modification of society, but rather in the congregation's determination to worship Christ in all things."[10]

I want to note the image of church which Hauerwas advocates in

9 Ibid., 43.

10 Ibid., 45.

this post-Christendom time: It is that of "an alternative polis" a "countercultural social structure called church."

> "It seeks to influence the world by being the church, that is, by being something the world is not and can never be, lacking the gift of faith and vision, which is ours in Christ. The confessing church seeks the visible church, a place, clearly visible to the world, in which people are faithful to their promises, love their enemies, tell the truth, honor the poor, suffer for righteousness, and thereby testify to the amazing community creating power of God. The confessing church has no interest in withdrawing from the world, but it is not surprised when its witness evokes hostility from the world."[11]

In other words, the best thing the church can do for the world is to become a vital and visible community of faith.

A stretch of more than forty years separates the perspectives of Niebuhr and Hauerwas. Niebuhr rejects a sharply conflictual view of the church and the world. Like Augustine who rejected the Manichean dualism of his times, Niebuhr does not see any "empire" which is entirely evil, which is without points of light, which is beyond conversion. And he does not see any institution (such as the church) which is without its shadow side. Herein lies the foundation and impetus for the transformation and conversion of culture.

However, Hauerwas sees the empire of America (modernity, the west) as beyond redemption . For him it is time to stop propping up the empire and to go about building an alternative culture of faith. His position here is not unlike that articulated by Alasdair MacIntyre at the end of his important book *After Virtue*. MacIntyre draws parallels between the decline of the Roman and American empires and says that the crucial point for the church occurred when men and women of good will ceased to identify with the status quo and turned aside from the task of shoring up the empire. They set themselves to "the construction of new forms of community within which the moral life could be sustained so that both morality and civility might survive the coming ages of barbarism and darkness."[12]

11 Ibid., 47.

12 Alasdair MacIntyre, *After Virtue*. Nortre Dame: University of Notre Dame Press, 1981. 244.

These are obviously two very different interpretations of the relationship between church and culture. Do these differences arise because of their differing historical contexts? Probably. The American empire is in a much different moment than it was in the confident years following World War II when the nation was developing into an empire. It has now entered a period of decline—a fact much discussed by historians and social scientists, and now an emerging topic in theological reflection. In 1951, the Christendom model was still very much in evidence in the western world—even (and perhaps especially) in the United States in which the church relied on the state to ensure freedom of religion. However, by 1991, the mainline churches were clearly experiencing the problematic realities of Christendom.

Yet, the differences between Niebuhr and Hauerwas arise from more than a difference in historical conditions. If Niebuhr is more willing to grant a measure of truth and goodness to the culture, he is also willing to acknowledge the illusion and sinfulness which can exist within the church or an "alternate culture." Hauerwas seems less aware of this and seems more intent on building a culture of resistance than in ensuring the possibility of prophetic critique within the church.

VACLAV HAVEL'S REFLECTIONS ON A "PARALLEL CULTURE"

It is instructive to compare and contrast the reflections of Hauerwas with those of Vaclav Havel, the former playwright and dissident now president of the Czech Republic. Havel is not a professional theologian but his reflections (most notably his Fourth of July address in Philadelphia) have been recognized as statements of great theological significance in the search for a meaningful human future in a time of an increasingly dehumanizing globalization.

I am interested in Havel because he too had to contend with a materialistic culture which he saw as seriously corrupted by a lie. What was the lie of communism? A lie not unlike our own. The lie is that you will be happy if you just have enough things. Like Hauerwas, Havel sees the importance of being able to live in a community marked by a more spiritual set of values. However, there are significant differences between these two writers.

Havel's key reflections on building an alternative to the dominant culture are to be found in a long essay entitled "Power and the

Powerless" which was circulated clandestinely in Czechoslovakia in 1978.[13] By that time, Havel was a well-known writer, one of the founding members of the Charter 77 movement (a human rights group). He had been arrested and imprisoned several times.

According to Havel, the people of his country and the eastern bloc were living in a post-totalitarian system, a system which was sustained by an ideology, by lies. "Ideology...pretends that the requirements of the system derive from the requirements of life. It is a world of appearances trying to pass for reality."[14] His description of this world of appearances sounds somewhat familiar and, indeed, Havel has consistently said that the materialistic culture of communism was not that different from the materialism of the west:

> Individuals need not believe all these mystifications, but they must behave as though they did, or they must at least tolerate them in silence, or get along well with those who work with them. For this reason, however, they must live within a lie. They need not accept the lie. It is enough for them to have accepted their life with it and in it. For by this very fact, individuals confirm the system, fulfill the system, make the system, are the system.[15]

The ideology of the system provides a kind of "metaphysical order" which gives the system a legitimacy and order. It is the glue binding it together, "the principal guarantee of the inner consistency of power."[16] If such is the case, then a system cannot be changed simply by changing the people in the power structure. If the ideology of a system remains untouched, then the power of a system remains even as particular individuals come and go. Havel is dubious about so-called political alternatives and about the possibility of changing the political system.

When the excuses or the lies (the ideology of the system) are accepted, then the power of the system is not merely "power over." The power of the system becomes inwardly constituted within the person. Human beings are not only victims of the system, they are also its instruments and supporters. The dividing line between the victims and supporters of

13 in Vaclav Havel, *Living in Truth*. London: Faber & Faber, 1987, 36–122. (edited by Jan Vadislav).

14 Ibid., 44.

15 Ibid., 45.

16 Ibid., 48.

the system does not lie "out there," but runs through and within each person.

The lies of the system have power only as long as people are willing to live within the lie. They will continue to do so, according to Havel, as long as they are seduced by consumerism.

> Is it not true that the far-reaching adaptability to living a lie and the effortless spread of social auto-totality have some connection with the general unwillingness of consumption-oriented people to sacrifice some material certainties for the sake of their own spiritual and moral integrity? With their willingness to surrender high values when faced with the trivializing temptations of modern civilization? With their vulnerability to the attractions of mass indifference?...And do we not in fact stand...as a kind of warning to the West, revealing to it its own latent tendencies.[17]

Havel's question, so similar to our own, is how we can live in truth in a culture of lies. Living in truth is a multi-leveled activity: it has an existential dimension (returning humanity to its inherent nature), a noetic dimension (revealing reality as it is), a moral dimension (setting an example for others), and an unambiguous political dimension. Havel's theological anthropology, as it were, is his conviction that, even in a culture of lies, truth exists as a "repressed alternative" or as a "hidden sphere." Indeed the culture of lies presupposes the desire for truth. "In its excusatory, chimerical rootedness in the human order, it is a response to nothing other than the human predisposition to truth."[18]

Just as the corruption of the system exists within each of us, so too does the alternative reside within us. The line between the dominant culture and its alternative lies within each of us.

Havel's strategy is to strengthen what he calls "the prepolitical," or "the existential," or "the independent sphere of life." He doubts that merely rearranging the political system or the political players will change the political and economic system, because the system is within each person; they are the system.

What is needed is a radical change within the human person: "Such a change will have to derive from human existence, from the fundamental reconstitution of the position of people in the world, their relation-

17 Ibid., 54.

18 Ibid., 57.

ships to themselves and to each other, and to the universe...If a better economic and political model is to be created, then perhaps more than ever before it must derive from profound existential and moral changes in society."[19]

Havel does not think that such a "conversion" can happen in a culture of lies. Nor does he think that it is possible to live entirely outside such a culture—neither geographically nor spiritually nor psychologically. A self-enclosed counter-culture does not seem to be an option for him. What he proposed instead is what he calls a "parallel culture." Such a culture creates a space within a dominant culture so that human beings can strengthen the alternative which exists within themselves.

It is in a "parallel culture" that one begins to experience one's identity as being primarily *for* the truth instead of being defined by what one is *against*, i.e. against the system of lies. Such a culture has a wide spectrum of expressions and activities: "self-education and thinking about the world, through free creative activity and communication to others, to the most varied free, civic attitudes, including instances of independent social self-organization."[20]

The people involved in such a parallel culture could be musicians, philosophers, historians, scholars, (participants in the "private universities"), publishers of underground periodicals (*samizdat*), teachers who teach young people things they would not learn in the state schools, religious people who practice their faith, people who try to put real meaning in trade unions, those who are not afraid to denounce injustice.

What unites all these activities and persons is the desire to strengthen the truth: "In other words, serving truth consistently, purposefully and articulately, and organizing this service."[21] This prepolitical activity will, according to Havel, provide a new ground for alternative political structures and movements.

And for one, brief shining moment it did seem as if this was possible. One day, the people who had been living in the truth in a parallel culture in Czechoslovakia (and elsewhere in the Soviet Union) began to live in the truth in a more public way. They began to walk in the streets with candles. They said that the emperor had no clothes. They said that it was a lie that they would be happy if they just had more things. They

19 Ibid., 70–71.

20 Ibid., 85.

21 Ibid., 87.

said that, as human beings, they needed faith and they needed hope and meaning and purpose to their lives.

And the system which had seemed so absolute and unchanging fell like a deck of cards. Once the lie no longer compelled universal assent, the power of the system was gone. The communist system collapsed—not because of the vast weaponry of the west but because its own citizens no longer believed in it.

Nevertheless, we know that the profound change which Havel and others had hoped for has not happened. There has been a vast reversion to materialism—only now with a far greater disparity between the rich and the poor. As one Christian activist from East Germany said sadly, "I thought that freedom was about more than shopping." Perhaps the "parallel culture" needed more time to grow and strengthen people in the truth.

However, Havel's description of the "parallel culture" which was built in the midst of a dominant socio-economic system is suggestive for us here and now. He talks about the importance of a parallel information network, a parallel system of education, parallel trade unions and some kind of parallel economy. What would it mean to begin to think about, to imagine, the church as a parallel culture—as the prepolitical ground of a new politics? What are the implications of imagining religious communities and theological schools as the seed and fruit of parallel cultures? What are the problems and the possibilities inherent in developing women's groups as parallel cultures within the dominant patriarchal culture and patriarchal church?

There is a subtle difference of tonality in the writings of Hauerwas and Havel. Hauerwas is blistering in his indictment of our corrupt culture and devastating in his critique of the church's collusion with the cultural powers that be. He seems to want to draw a line between that dominant culture and the church as an alternative culture. Like Noah, he seems to want to construct an ark so that some alternative may be possible as the flood waters of these times rise. It is difficult to see how Hauerwas can escape Niebuhr's critique of the dualism which lurks in countercultural models of the church i.e. the darkness in the culture without and the light within the Christian community.

Yet, it seems important to recall that the gospels tell us "God so loved the world." Any resistance to the dominant culture, any creation of an alternative, should ultimately be for the sake of the world which is so loved by God. Vaclav Havel and, indeed, someone like Margaret Brennan, seem to have a deeper sense of the "universal" dimension of a particular counterculture.

> "It is not something partial, accessible only to a restricted community, and not transferable to any other. On the contrary, it must be potentially accessible to everyone; it must foreshadow and general solution and, thus, it is not just the expression of an introverted, self-contained responsibility that individuals have for themselves alone, but responsibility to and for the *world.*[22]

Havel's perceptive analysis of how and why human beings internalize the values and patterns of the dominant culture leads him to conclude that a strict separation from this culture is not possible. And he also seems to suggest that it is not a desirable separation. The parallel culture can and should reenter the dominant culture. Yet, if the experience of the attempts to change the socio-economic system of communism (which is now also neo-liberal) is indicative, the question of *timing* becomes an important point of discernment. How long do we need to live in the truth before we are ready to confront the lies of a culture? Answering this question involves the integration the wisdom of the tradition of spiritual discernment and the practical judgment of political decision.

22 Ibid., 103.

Spirituality
and Social Movements
in the Post-Modern World

Lee Cormie

Introduction

"From the perspective of our faith we see that peoples' movements necessarily are the means by which we serve the Kingdom," affirmed more than one hundred pastors and theologians who signed the *Kairos Central America* document.[1] This is one of the most fundamental insights of the "new" voices in theology. The Spirit of Life is incarnated in society and history through social movements which nurture the dignity of marginalized peoples, and their capacity to speak for themselves about their experiences and hopes—and about the concerns of the earth and of peace—and to take the initiative in reaching out to others in broader solidarity, and to have a voice in the centers of decision-making power.

In individualistic cultures like ours in Canada and the United States though, it has been especially difficult to (re)establish the "communal" and "social" dimensions of spirituality, and its "active," "historical," even "political" character. No one has labored in these fields more faithfully than Margaret Brennan. In this spirit, I offer the following partial and provisional probes into the winding trajectories of popular movements in the struggles for justice, peace and the integrity of creation since the 1960s; the changing contours of a rapidly "globalizing" and "post-modern" world; and the path of the Spirit beckoning us into the future.

Irruptions of Popular Movements

Spirituality concerns the ongoing process of disciplining—ordering and reordering—the whole of our lives in our quest to follow the Spirit along the twisting paths of history. It concerns the concrete ways we

1 *Kairos Central America*, signed by a group of more than one hundred priests, pastors, theologians, and lay leaders from Central America, English translation New York CIRCUS, *Amanecer* (English Edition), No. 3, June, 1988, para. 68.2.2; cf. also Gutiérrez, Gustavo, *The Power of the Poor in History*, trans. Barr, Robert, Maryknoll, NY, Orbis Books, 1983.

each forge for living out our hope and faith in this life on earth. Of course, it concerns intensely personal matters.

But personal horizons, challenges, and choices are also profoundly shaped—and *mis*shaped—by "social" factors which largely escape notice in liberal (and neoliberal and neoconservative) perspectives. And the "new" voices which have erupted in social movements and in theology since the 1960s have challenged this blindness. In some ways, like modern liberalism, they have witnessed to the great faith in progress, in the possibilities for fulfillment in history. But in different ways they have also criticized liberal thought, stressing the economic, cultural, and political obstacles to progress in modern society, and more generally, the significance of social structures and dynamics shaping every dimension of our personal destinies, even the innermost recesses of our psyches and our souls. And they have labored to renew spirituality and politics by bridging the gaps in the ways we frame the personal and the social, the sacred and the secular, gaps still plaguing mainstream expressions of politics and economics, and mainstream Christian expressions of hope and faith.[2]

In the decades since the 1960s we have been graced with so many witnesses to the Spirit: Third World liberation movements, the civil rights and black power movements, the feminist movement, Third World solidarity movements, indigenous movements, the labor movement, community organizations, the peace movement, gay and lesbian movements, the ecology movement seeking to give voice to other species and the living earth itself, and a growing number of others seeking to give expression to the "mixed" identities of hyphenated Canadians and Americans, and of immigrants everywhere.[3]

Resonating in this respect with older "conservative" reservations about the wonders of "liberal" society and the global capitalist order, they rejected the claim that "modern," "Western" society was the highest stage of development where the major social barriers to equality had

[2] This gap is not a problem for fundamentalist and new right Christians; in this respect they share with "radical" Christians suspicion of conventional "liberal" expressions of Christianity.

[3] Concerning the particular genius of Hispanic Americans as the heirs and architects of a broader, more inclusive, "mixed" culture, cf. Elizondo, Virgil, *The Future Is Mestizo: Life Where Cultures Meet*, Bloomington, Ind., Meyer-Stone Books, 1988.

been overcome and each individual's fate was in his or her own hands. In different ways they challenged the confidence that sufficient progress in overcoming remaining injustice could come—was coming automatically—through business as usual leading to economic growth, increasing incomes, and modest reforms. They questioned the arrogance of the guardians of this sacred order, and the forms of authority, accounting frameworks, and decision-making processes which excluded the voices and concerns of the marginalized. And they questioned the ideologies legitimating their exclusion, and every god who was appealed to in blessing this unjust order.

These "new" voices insisted that, despite the rhetoric of equality and freedom, there were substantial groups with little or no voice in defining the reigning canons of beauty, goodness, and holiness, in articulating the meaning of history, in setting goals and priorities, and most importantly, in making decisions in every sphere. In many families, and in the churches, too, these "others" were designated as inferior, and abused by those who believed themselves superior. They were systematically exploited in the marketplace, and in the informal sector outside the market as well. And everywhere they were barred from positions of power. In the end these "others" constituted the great majorities around the world, including in the "advanced," "modern," and "wealthy" societies. Individually and together they raised radical questions concerning how our most sacred beliefs, traditions, institutions and authorities could have permitted, or worse encouraged, such blindness, inequality, and suffering. They challenged the ways we had all learned to see and feel about ourselves and others, our history and our hopes. And in different ways they called for changes that go to the roots, "radical" changes. Blacks, women, and indigenous peoples irrupted, breaking into the debates about our history, our future, and the conflicting spirits calling for our devotion; and gradually Hispanics, Asian Americans and Asian Canadians and other "others" added their voices. Through exiles and refugees, various solidarity groups, and returned missionaries, Latin Americans, Africans and Asians added their voices to the debates in North America. Swept up without a voice in the economic, political and cultural changes sweeping the world, students and activists in local communities (re)asserted their right to participate in articulating goals, setting priorities and making the decisions that affected them. Gays and lesbians spoke up concerning the crippling straightjacket of the reigning expressions of heterosexuality. Here and there workers called attention to workplace health and safety hazards, automation and speedups, overtime for a few and mass layoffs for many, unemployment, the explosion of part-time jobs, and underemployment. Convinced that concerns with non-violent conflict resolution and peace had been marginalized altogether, especially in the agendas of the Reagan and Thatcher governments and their supporters, peace activists added their voices denouncing the worldwide militarism resulting in the deaths of millions, increasing insecurity and distorted development everywhere. And convinced that the voice of the earth—of other species, the biosphere as a whole, and the foundations of all for all species, including humans—had long been silenced, environmentalists added their voices to those of indigenous peoples and many farmers calling attention to ecological crises and the urgent need for change in industrial culture, corporate and government policy making.

Surprising all who saw the churches as bastions of hopelessly "conservative" cultures and values, these new voices quickly found expression in the "new" voices in spirituality and theology. In theological terms, in Catholic circles, against the background of an ethos which viewed history in essentially unchanging terms and almost reflexively endorsed the status quo while virulently condemning all proponents of radical social change (uncritically lumping together an extraordinary diversity of Marxists and socialists and other critics of the status quo), these new ways of doing theology reasserted the salvific significance of society, politics, and history. They insisted on the option for poor and oppressed peoples, and later for the earth too, as God's option. They launched a renewed critical examination of "conservative" apologies for established authorities and institutions, and containment of hope for poor and working people everywhere, and for the earth. And they pressed for the renewal of hierarchical structures and forms of authority in the church so that the whole people of God could witness concretely to God's promise of dignity and the fullness of life for all in the context of the promises of new technologies and forms of social organization.[4]

In mainstream Protestant circles, the challenges of these movements

[4] For a recent popular expression of this spirituality from Latin America, cf. Casaldáliga, Pedro and Vigil, José-María, *Political Holiness: a Spirituality, of Liberation*, trans. Burns, Paul and McDonagh, Francis, Maryknoll, NY, Orbis Books, 1994.

were somewhat different. By the end of the 19th century liberal Protestants had incorporated into theology the notions of historical change and progress, and developed the new discipline of social ethics for addressing the questions of social order and change. But the "new" liberation voices also challenged this form of "social" concern. These new voices took their starting points from the sufferings and hopes of marginalized groups at home and abroad. They rejected "liberal" notions of history as "progress," culminating in the ideals of European and North American "middle class," white, male affluence, privilege, and power. And they challenged the churches to rediscover the prophetic tradition in the bible, to abandon the naive hope in incremental "reforms," to pressure for more fundamental changes, to witness concretely to nothing less than the coming of the reign of God in history.[5]

DEATH(S) OF HOPE(S)

As the 1990s unfolded, though, there was no doubt that these movements were in crisis. And the concrete expressions of hope incarnated in them were dying, or dead.

We have witnessed the death of the "radical" ("socialist" and "communist") expressions of hope born in the 19th century, for global social changes and great leaps forward in the historic quest to overcome class conflict and imperialism; this historic end was symbolized in the tearing down of the Berlin wall in 1989.

We have witnessed the death of the hopes of the "new social movements" of the 1960s and 1970s for great leaps forward in overcoming racism, patriarchy, militarism and systemic reliance on violence in maintaining "order," heterosexism and homophobia, and ecological devastation in the wake of industrial development everywhere.

Unfolding in different ways in different contexts since the mid-1970s (and even before in some contexts), the organizations and movements which incarnated these hopes have been blocked, their agendas reduced to defending earlier gains, and their supporters scattered, or simply

5 Cf. for example Bonino, José Míguez, *Toward a Christian Political Ethics*, Philadelphia, PA, Fortress Press, 1983; and Harrison, Beverly Wildung, "Agendas for a New Theological Ethic," in Tabb, William, ed., *Churches in Struggle: Liberation Theologies and Social Change in North America*, New York, NY, Monthly Review Press, *1896*, 89–98.

crushed. These historic defeats are spiritual as well as political. As Nicaraguan theologian José María Vigil has insisted, from the point of view of the South, the triumph of capitalism is a "complete defeat, a defeat not only of society but also of human nature and, ultimately, of God."[6] We have heard similar cries haunting "affluent" countries like Canada and the United States, and increasingly rising throughout the former "Second World" as well. (At the same time, in the eyes of faith at least, there have been signs of the rebirth of these hopes, especially in the remnants and reincarnations of these movements and the broader circles influenced by them; we will return to them below.)

Neoliberal and neoconservative spokespeople simply dismiss these voices, claiming that in different ways all were fundamentally mistaken about human nature and modern social order, and about the nature of the problems facing their constituents and the world. In their view, these expressions of hope were rightly destined for the dust bin of history.

I would like to suggest, though, that this interpretation is far too superficial and self-serving.

Of course, as prominent spokespeople within each have acknowledged, none of the progressive movements of the last one hundred years was perfect. At times, some voices were abrasive, shocking, and self-serving. And sometimes prominent leaders and spokespeople, at least those trumpeted in the mainstream media, had an exaggerated sense of their authority and power. Clearly, these movements and their leaders were not immune to human frailties and temptations. But open dialogue, criticism, and debate have been prominent features of each movement. In any case, mistakes, sins, and limitations do not provide a basis for distinguishing between them and other movements and organizations. And focusing on them distracts attention from a host of fundamental issues involved in their rise and fall.

It is also true that different currents emerged within each movement. These plural currents reflected the great wealth and diversity of experience and concerns within each, and their dynamic, evolving char-

6 Of course, Vigil continued, "we do not accept this failure as definitive, because we do not accept the failure of God." Vigil, José María, "What Remains of the Option for the Poor?," *LADOC*, 25, 2, 1994, 5. Concerning Africa, cf. M'Mwereria, Godfrey, *The Root Causes of Debt Crisis in Africa: an Agenda for Continental Collective Self-Reliance*, Nairobi, Kenya, Southern Network for Development, Africa Region, nd.

acter. At times, though, these differences also manifested significantly different analyses and visions, moving in multiple, even conflicting (sometimes simultaneously "conservative," "reformist," and "radical") directions at the same time. And numerous major unresolved questions remained concerning the nature of the "system" (class-divided nation, imperialist world system, patriarchal, ethnocentric, and/or ecologically hostile "western" and/or "Christian" civilization), the character and scope of the desired alternative(s) to it, and strategies for moving forward. In the end none of these movements was able to offer an encompassing and coherent vision of the destiny to which all are called, complete analysis of the failures of the existing order, or a magic formula for transforming history. And collectively they were far from achieving a common voice, shared vision, broader solidarity, and shared strategies.

These limitations weakened these movements and their impacts. And, more generally, they contributed to the increasingly generalized sense of crisis haunting existing versions of revelation, reason and science, the movements they inspired, and the institutional forms in which they were incarnated. They contributed to the growing confusion about the possibility of ever again speaking authoritatively—with a single voice—in the name of large communities, or "society" as a whole, never mind the whole world. They contributed to the seemingly endless pluralism, confusion, and paralysis marking our "post-modern" world.

But this can not be the last word either. For these movements were not the only, or major, sources of the growing crisis of authority in every realm (as we will see below). And, more generally, it is vital to attempt to situate the stories of these movements in a broader historical context. Surely we are still at the early stages of emergent global dialogues, being made possible for the first time in history by advances in technologies, social organization, communication and transportation. In this light, we can recognize as truly historic the irruptions of those whose voices were previously distorted and silenced in public forums and centers of power. And we can affirm the possibility—indeed necessity—of new, ever more inclusive dialogues which alone promise true hope in nurturing wider understanding, broader solidarities, and more effective collaboration in addressing the challenges confronting the world.

These movements have helped to give voice to the experiences, concerns, and hopes of countless marginalized peoples, and to support their having a voice in the debates over the future. They have witnessed to the enduring power of the Spirit to inspire the transformation of our lives, social order, and the course of history. And it is in their remnants and reincarnations today, I am deeply convinced, that we find the seeds of hope for a different future.

Simple dismissals of them, then, seriously distort our interpretation of realities and challenges confronting us at the end of the millennium, and the possibilities of hope for the future. They also overlook the deep crises which have overtaken "liberalism" and "conservatism."

For born-again liberals—"neoliberals"—have repudiated the hope for "development" and "progress" which was born in the 19th century and matured after World War II, defining the welfare state in the First World and the possibilities for "development" in the Third World (and promised to the "Second World" too, if only their peoples would bless "private property" and "free markets").

And, ironically, born-again conservatives—"neoconservatives"—have proclaimed most loudly the "revolutionary" character of their agenda. For at least since the 19th century conservatives have been the most insistent on the fallen and sinful character of human nature, the most concerned about the fragile character of social order, the most emphatic about the importance of "traditional" values, authorities, and institutions, the most skeptical about utopian dreams to perfect society and liberals' and radicals' faith in their ability to manage large-scale social change. By the 1980s though, neoconservatives were joining neoliberals around the world in championing "revolutionary" changes in the social order that had been forged in the context of over one hundred years of great turmoil, suffering, and conflict.

In social terms, we are witnessing the end of a civilizational epoch, the great historical project of the last one hundred years and longer of liberal capitalism, which has oscillated between crisis, reform, and crisis, and of the efforts of its many critics to forge "radical" alternatives to it (some of which were quite "conservative"), which also helped to shape its trajectory. This is increasingly obvious in the "developed," "affluent" countries like Canada and the United States. Our country is "on the edge of disintegration."[7] The policies of corporations, governments, and the international financial institutions—the World Bank, the International Monetary Fund (IMF), and the World Trade Organization

7 Laxer, James, *False God: How the Globalization Myth Has Impoverished Canada*, Toronto, ON, Lester Publishing Limited, 1993, 3.

(WTO)—are everywhere promoting increasing underemployment and unemployment, deepening poverty, the destruction of communities, alienation, racism, pathology, and violence on the streets and in homes. It seems "as if we were all taking part in some kind of assisted national suicide."[8]

In ecological terms, we are witnessing the end of an evolutionary epoch. Tens of thousands of species on earth are becoming extinct, at a rate unsurpassed in the history of life on earth. And in the midst of deepening ecological crises, the future of many tens of thousands more, including the human species, is in question.

At the end of the second millennium since the birth of Jesus, we are witnessing the death of the dominant modern—"conservative," "liberal," and "radical"—expressions of hope in history.

And more than ever we—and tens of thousands of other species—lust for the resurrection of hope in history.

Global Orthodoxy

Strangely, the prophets of endless post-modern pluralism and relativism have overlooked the establishment of orthodoxy on a scale historically unprecedented in its global reach and power. In the centers of power in the "new world order" there is only one legitimate discourse for discussing the world's problems—neoclassical "economics." There is only one path to salvation, for all people, no matter what their own traditions and values, and historical starting points—East or West, North or South, rich or poor.[9] And in the decisions of policy makers in corporations, governments, and the international financial institutions, this orthodoxy is transforming "reality" in countless ways, incarnating its structural ordering principles and priorities in the very fabric of life on earth.

In a positivist spirit, the high priests of this project of restructuring life on earth argue that it is the only option, that all others have been condemned in the judgment of history. But all along there has been a

8 Walkom, Thomas, "Silent Night, Holy Night, All's Not Calm, All Is Blight," *Toronto Star*, 23 December, 1992, A25. Concerning the crisis in the United States, cf. for example Greider, William, *Who Will Tell the People: the Betrayal of American Democracy*, New York, NY, Simon & Schuster, 1992.

9 Cf. Mihevc, John, *The Market Tells Them So: the World Bank and Economic Fundamentalism in Africa*, Penang, Malaysia, Third World Network, 1995.

growing chorus of other voices offering other insights, analyses and judgments, and proposing different paths forward; this chorus reflects the "new" voices referred to above and many more traditional "liberals" and "conservatives" too, along with those from other cultural and political traditions altogether. The triumph of neoliberal orthodoxy, then, can not be explained as the result of consensus among scholarly experts at the end of a process of a civil, rational, open, debate conducted according to scientific canons concerning appropriate data, theoretical frameworks, methods, hermeneutics, analyses, and judgments. This triumph has owed much to power.[10] And this power requires deafness to the cries of so many others, blindness to the destructive effects of its policies, and the narrowing of the sphere of democratic dialogue and debate.

Everywhere, critics across a broad range of perspectives are pointing out, policies in the name of the market god are producing "development" benefiting only a few. They are multiplying the incomes and wealth of the already privileged, extending and reinforcing their power, and burdening the victims with the cost of this "restructuring"—unemployed, under employed, and ordinary working people, increasingly many in the "middle classes" too, the poor, the sick, children and the elderly especially, other living beings, and the delicate dynamic of the biosphere itself. Everywhere they are increasing the gaps between rich and poor, both among countries and within them, eroding local communities, families, and whole nations, and undermining the capacity for civil dialogue and debate. And everywhere, through sins of commission and omission, they are contributing to environmental devastation and species extinctions.

At this early stage of global dialogues, there is little agreement about the path forward, or about how to provision ourselves for what promises to be a long and hazardous journey into the future. However, in the reports from a growing number of other communities, centers, organizations, and networks, it is clear that the path of the Spirit of life leads—must lead—in a fundamentally different direction.

[10] Neoliberal/neoconservative ascendance was engineered largely without widespread popular support, indeed in opposition to public opinion on many basic matters; cf. Ferguson, Thomas and Rogers, Joel, *Right Turn: the Decline of the Democrats and the Future of American Politics*, New York, NY, Hill and Wang, 1986; and Warnock, John, *Free Trade and the New Right Agenda*, Vancouver, BC, New Star Books, 1988.

Epochal Changes

Deeper insight into the challenges before us at the end of the 20th century requires attention to other important factors in the unraveling of established forms of order and hope. We are at still an early stage in their development and in our efforts to grasp their scope and scale, but it is increasingly difficult to avoid the conclusion that we are living through epochal transformations in the nature of life on earth.

For the first time in history, global media and transportation systems are concretely linking people around the world in more than a spiritual and genetic sense. Global institutions promise the possibility of responding in new ways to the needs of poor and marginalized peoples, and to the ecologically devastating effects of industrial development. Genetic engineering is leading to the alteration of species, the creation of literally new life forms, and the alteration of the dynamics of evolution. And these developments are transforming the scope of human power, responsibility, and freedom.

Of course, the potential for evil is growing too, forging ever greater concentrations of power and social structures which marginalize the great majorities and their concerns; wrecking ecological devastation from deep in the earth to high in the heavens; spewing violence and the seeds of violence everywhere; mis-stepping along the hazardous, narrow, and dark path of genetic engineering.

Whatever the outcomes though, it seems clear that human nature, the nature of society in all its dimensions, "nature" itself, the dynamics of evolution, and the course of historical change are all being radically transformed. Perhaps only appeal to religious images can suggest the magnitude of these changes. I find myself continually driven back to the story of creation in the Book of Genesis, which has long framed Christian, and Jewish, reflections on the parameters and meaning of human existence.

But in contrast to the images in Genesis of our primordial parents, Adam and Eve, humanity in the closing decades of the 20th century is less and less a passive actor confronting an already created, God-given natural world. Humans—better *some* humans—have their hands on levers of power which used to be exclusively in God's hands. More truly than ever before in history, created beings have become co-creators, along with the Creator (or the "laws of nature" or "evolution"), in shaping—and *mis*shaping—the course of life on earth, and sharing responsibility for it. Our relationships to creation and to the Creator—perhaps the Creator too—are being fundamentally transformed.[11] And we seem poised on the brink of nothing less than a new creation.

Of course, these prospects may inspire both overwhelming despair and unbounded hope, whose winds sweep our souls in different wrenching combinations. And we are still at the early stages of crafting concrete hope-filled responses to these new possibilities and responsibilities. There is no doubt, though, that these changes have overwhelmed familiar formulations of hope and faith, and the theoretical frameworks, ideologies, political projects, social structures, and institutional forms which incarnated them.

Rebirth(s) of Hope(s)

On the eve of the third millennium since the birth of Christ, the renewal of hope requires nothing less than reinventing "civilization"—cultures and identities, religions and spiritualities, institutions and social structures, ethics and politics—in the context of epochal changes and of a sustained struggle for hearts and minds against the greatest concentrations of power in history. There are no blueprints. But as I have participated in discussions in various movement circles, and listened in on conversations and debates in others, and in the broader circles inspired and challenged by them, I have heard a growing number of reports of convergence on some basic points of reference.[12] At this early stage, dialogues are still generally dispersed and fragmented; the institutional capacity to sustain them is underdeveloped; and there is no agreed-upon

11 At the very least, these developments pose basic questions for our images, concepts and frameworks for thinking about the Creator and creation. As the recent working document on ecology of the Canadian bishops states, the symptoms of ecological crisis point to an unresolved issue for theology concerning the relationship "between God's transcendence and God's immanence." Episcopal Commission for Social Affairs of the Canadian Conference of Bishops, *The Environmental Crisis: the Place of the Human Being in the Cosmos*, Ottawa, ON, Canadian Conference of Catholic Bishops, 1996, 38.

12 Cf. for example Ruether, Rosemary Radford, "Spirituality and Justice: Popular Church Movements in the United States," in Bianchi, Eugene and Ruether, Rosemary Radford, eds., *A Democratic Catholic Church: the*

formulation of these insights. For many though, seeing with the eyes of faith, this convergence is a sign of (re)emergent hope. Here I can only sketch what I see as five constitutive dimensions of this hope.

(i) The option for the poor and marginalized remains central. But the understanding of this central axis is being continually refined and developed in complex ways, especially in response to the "new" faces of the poor and their uncovering of the dynamics of ethnicity and race, gender and sexual orientation, as well as class and national origin, in shaping social order, along with the crosscutting "new" concerns of peace and ecology. This development is always "universal" in principal, excluding no one, including the rich, from the call to conversion. But it is also clear in fundamentally reconfiguring our conversations, research and education, liturgies and politics to include centrally the voices of the previously marginalized. And it is evident concretely in emerging broader solidarities, and in networks, coalitions, and movements bridging historic gaps and barriers in shared struggles for a different future.

(ii) These efforts are marked by a new kind of "epistemological humility" in the "post-modern" world. There is a renewed awareness of the inevitably partial and limited character of all human knowing, and a corresponding shift from the priority of abstract "theory" over concrete practice to "praxis" (theory, analysis, and practice in systematic, ongoing, self-conscious interaction). These approaches are more self-consciously contextual and grounded in the experiences and hopes of specific communities and movements. They are more "reflexive," in contrast to positivist hermeneutics, applying to themselves, schools of thought, and movements the same descriptive and explanatory apparatus applied to others.[13] And they are more "popular" and inclusive, emphasizing widespread participation, "pluralism," and "democracy" in the development of theoretical frameworks and analyses, research and education.[14]

Reconstruction of Roman Catholicism, New York, NY, Crossroad, 1993; 189–206; and Lind, Christopher and Mihevc, Joe, eds., *Coalitions for Justice*, Ottawa, ON, Novalis, 1994.

[13] Cf. for example Harding, Sandra, "Common Causes: Toward a Reflexive Feminist Theory," *Women & Politics*, 3, 4, 1984, 27–42.

[14] Cf. Sheth, D. L., "Alternative Development as Political Practice," in Mendlovitz, Saul and Walker, R. B. J., eds., *Towards a Just World Peace: Perspectives from Social Movements*, London, Butterworths, 1987, 235–251; and Wainwright, Hilary, "A New Kind of Knowledge for a New Kind of State,"

(iii) At the heart of these efforts is a rejection of conventional notions of "progress" and "development,"[15] and a redefinition of "economics" in ways which explicitly include religio-cultural, moral, and political concerns, in relation to specific issues and in the development of economic discourse in general. Concretely, this means that the most basic categories of economics and accounting must be revised in order to include the concerns, insights, and points of reference which are invisible in conventional frameworks, or are regarded as "external," but which have emerged as central in the experience of these movements (like culture and spirituality, gender, race and nature, violence and militarism, as well as class and world system).[16]

(iv) In an increasingly global system, it is clear that "Third World" voices, and the voices of marginalized cultures within the First and Second Worlds, are indispensable in these efforts. In particular, it is clear that "poverty," and alternatives to the dominant culture of consumerism, must be at the center of all plausible visions.

Wolfgang Sachs' insight is being echoed widely: "If all countries 'successfully' followed the industrial example, five or six planets would be needed to serve as mines and waste dumps. It is thus obvious that the 'advanced' societies are no model; rather they are most likely to be seen in the end as an aberration in the course of history."[17] And another hope for the future is emerging—must emerge—to replace established versions of "development."

In this spirit, for the sake of the earth and the poor of the earth and ultimately of all, martyred Salvadoran liberation theologian Ignacio Ellacuría proposed the vision of a "civilization of poverty."[18] In nurtur-

---

in Albo, Gregory; Langille, David; and Panitch, Leo, eds., *A Different Kind of State? Popular Power and Democratic Administration*, Toronto, ON, Oxford University Press, 1993, 112–121.

15 Cf. Esteva, Gustavo, "Development," in Sachs, Wolfgang, ed., *The Development Dictionary: A Guide to Knowledge as Power*, London, England, Zed Books, 1992, 6–25.

16 Cf. Cobb, Clifford and Halstead, Ted, *The Genuine Progress Indicator: Summary of Data and Methodology*, San Francisco, CA, Redefining Progress, 1994.

17 Sachs, Wolfgang, "Introduction," in Sachs, ed., *The Development Dictionary*, 2.

18 Ellacuría, Ignacio, "Utopia and Prophecy in Latin America," in Ellacuría, Ignacio and Sobrino, Jon, eds., *Mysterium Liberationis: Fundamental Concepts*

ing the spirit of this new civilization, Latin Americans have much to offer the world, especially to those radically challenged to conversion in societies like ours most structured around consumerism. In the words of his colleague Jon Sobrino, "utopia in the world of today cannot be other than the 'civilization of poverty,' all sharing austerely the resources of the world so that they are sufficient for all."[19] And around the world others are coming to similar conclusions concerning the necessity, inevitability—indeed desirability—of a civilization of "austerity," or "subsistence," or "sufficiency."[20]

(v) Finally, an alternative civilizational project requires articulating anew the grounds for faith in the historical possibilities of hope for a future different from the abyss into which current constellations of power are rushing us all. Surely, for us Christians especially, the new voices which have found expression in the church have revealed once again the power of faith in society and history, and the capacity of a very old Judeo-Christian tradition to nurture new historical expressions of the "good news" in the midst of epochal change, and the rebirth of the church itself as a community of communities witnessing concretely to hope in history.

At the Heart of the Bible

In exploring the deeper significance of these movements, it is important to recognize how profound are the re-readings of history they have inspired, how often the central role of poor and oppressed people in nurturing hope for justice and peace has been confirmed, and how powerfully inspiring in the present these witnesses from the past have been.[21]

---

of Liberation Theology, trans. Brockman, James, Maryknoll, NY, Orbis Books, 1993, 314–315.

19 Sobrino, Jon, "Crucified Peoples, the Present Suffering Servant of Yahweh," *LADOC*, 1992, 17.

20 Cf. also Mies, Maria, "The Need for a New Vision: The Subsistence Perspective," in Mies, Maria and Shiva, Vandana, *Ecofeminism*, London, UK, Zed Books, 1993, 297–324; and Latouche, Serge, *In the Wake of the Affluent Society: An Exploration of Post-Development*, trans. O'Connor, Martin and Arnoux, Rosemary, London, England, Zed Books, 1993.

21 There is a growing list of this kind of critical rereading of history, its theological, missiological, and political implications. For example: Wilmore,

Especially important for us Christians, these new voices have inspired re-readings of the bible which have uncovered these revelations at its very heart.[22]

God's option for poor and oppressed peoples, and for creation, and the call to follow the Spirit of justice, peace, and the integrity of creation in solidarity with the poor, are central themes in the Hebrew and Christian scriptures. Of course, there is also ample testimony concerning the human condition, which is marked by frailties, ignorance, and sin. There is ample testimony, too, to the ease with which doctrines and institutions can become sacralizations of the power and hope of dominant groups, even in synagogues and churches. Throughout, these sins are portrayed as alienating and oppressive for the majorities, and in the end for the wealthy and powerful too, since they inevitably undermine social foundations and lead to chaos and/or defeat at the hands of enemies. Throughout, the bible confirms that struggles over "economics" and "politics" are also "spiritual" struggles over hope(s) and faith(s), between following idols or the Spirit of the true God.

The story of the liberation of the Jews from Egyptian slavery at the founding of Israel sets the stage for the revelation of God in the bible as the sole high God, the Creator and Redeemer of the cosmos, and liberator of the poor.[23] The stories of the prophets and of renewal movements

---

Gayraud, *Black Religion and Black Radicalism: An Interpretation of the Religious History of Afro-American People*, rev. 2nd ed., Maryknoll, NY, Orbis Books, 1983; originally published by Anchor Press/Doubleday in 1973; Fiorenza, Elisabeth Schüssler, *In Memory of Her: A Feminist Theological Reconstruction of Christian Origins*, New York, NY, Crossroad Publishing Co., 1983; Dussel, Enrique, *The Invention of the Americas: Eclipse of "the Other" and the Myth of Modernity*, trans. Barber, Michael, New York, NY, Continuum, 1995.

22 Elsewhere I have sketched the different but in important respects convergent insights of Latin American, feminist and African American theologians in rereading the bible; cf. my "Revolutions in Reading the Bible," in Day, Peggy; Jobling, David; and Sheppard, Gerald, eds., *The Bible and the Politics of Exegesis*, New York, NY, Pilgrim Press, 1991, 173–193.

23 Gottwald's *The Tribes of Yahweh* was ground-breaking in clarifying the social context and dimensions of emergent Israelite faith. Cf. Gottwald, Norman, *The Tribes of Yahweh: A Sociology of the Religion of Liberated Israel, 1250–1050 B.C.E.*, Maryknoll, NY, Orbis Books, 1979. Cf. also Pixley, George, *Exodus: A Liberation Perspective*, Trans. Barr, Robert, Maryknoll, NY, Orbis Books, 1987.

within Judaism reaffirm this Spirit again and again, in resistance to foreign domination, and often against the distorted forms taken by Israel and its religious leaders in shifting historical circumstances.[24] And in the story of the prophet Jesus who announced good news to the poor, and of the movement which formed around him, the Christian scriptures reaffirm the powerful presence of this Spirit in the world, in opposition to the false gods preached in elite expressions of Jewish faith and imperial Roman state religion.[25]

In rediscovering this Spirit in the bible, the "new" voices in the Church have grounded their often radical challenges to conventional modern expressions of Christian faith and spirituality, ecclesial organization, and pastoral practice, in the very heart of the tradition. In so doing they have uncovered deep well springs nurturing renewed faith in the face of great obstacles, profound humility and openness to new challenges of the Spirit, and joyful readiness for new departures. And in the process they have established themselves as more authentically "traditional" than both "liberal" and "conservative" expressions of Christian hope and faith.

They have rooted us more deeply in the past.

They are helping us to see more clearly the path of the Spirit into the future.

And more than ever we—and tens of thousands of other species—lust for the resurrection of hope in new and renewed social movements incarnating broader solidarities, more effective collaboration, and greater faithfulness to the Spirit in transforming history.

24 For a general introduction to the exegesis of the Hebrew scriptures reflecting these concerns, cf. Ceresko, Anthony, *Introduction to the Old Testament: A Liberation Perspective*, Maryknoll, NY, Orbis Books, 1992.

25 Concerning the Jesus movement, cf. Horsley, Richard, *Sociology and the Jesus Movement*, New York, NY, Crossroad Publishing Co., 1989.

Social Justice *and* Ecological Responsibility *as* Challenge *to* Integration *and* Conversion

Amata Miller, IHM

Thinkers from many disciplines have now concluded that if we are to be faithful disciples of the God who has formed and called us, the new probings and understandings in science, ecology and cosmology must reshape our understanding of who God is, who we are, and what our relationships with others and the world around us are/should be.

This paper is a reflection on how our living out of the new understandings of our place in the universe is linked with our gospel commitment to justice, especially for the economically poor.

- How are justice and ecological responsibility related to one another?
- What is required to act both in ecologically responsible *and* socially just ways in our world? Are there current examples of people doing this?
- How can we be agents of transformation toward more just and sustainable societies, toward a world whose institutions promote rather than hinder social justice and healthy environmental conditions?

The probing of these questions will proceed through a discussion of how justice and ecological responsibility are integrally linked—in the developing social teaching of the Church, in current realities and emerging alternatives, and in the action called for to bring about a just and sustainable world for all peoples.

Justice and Ecology in the Church's developing Social Teaching

The teaching of the Church on justice is dynamic and continually developing as societal realities and our understandings of them change. As ecological awareness and knowledge has expanded, this social teaching has included ecological responsibility in its reflection, analysis, and moral message.

Justice in the Church's Teaching

Central to the biblical tradition in both testaments is the teaching that the justice of a community is measured by its treatment of the powerless, named repeatedly as the widows, the orphans and the strangers in the land.[1] Fidelity to the God of the covenant, to the God of Jesus Christ, means participating in creating societies characterized by justice as well as compassion toward the neighbor whom God has called to be "a sharer on a par with ourselves in the banquet of life to which all are equally invited by God."[2]

John Paul II, in his first encyclical, *Redemptor Hominis*, identified a theme to which he would frequently return, the "chasm" between the living standards of rich and poor nations. He drew attention to the commercial, financial, economic and political structures which generate this gross inequality and he expressed his judgment that the current structures undergirding the world order are incapable of remedying the injustice which they have created.[3]

As the pontiff's global experience deepened his awareness of global poverty and its causes, he began to speak of moral *priorities*:

> "The needs of the poor take priority over the desires of the rich; the rights of workers over the maximization of profits; the preservation of the environment over uncontrolled industrial expansion; production to meet social needs over production for military purposes."[4]

This idea that some good things are more important than others, and must take precedence over them, is a new kind of moral teaching, strange to the ears of Catholics whose consciences were formed to focus on choices between good and evil. But, John Paul says, the depth of human misery is so widespread in our age, the injustice so flagrant, that

1 National Conference of Catholic Bishops. *Economic Justice For All: Pastoral Letter on Catholic Social Teaching and the U.S. Economy* (Washington, DC: U.S. Catholic Conference, 1986), Art. 38. (Hereafter cited as *EJA*.)

2 John Paul II, *Sollicitudo Rei Socialis* (December 30, 1987) published in *Origins*, 17:38 (March 3, 1988), Art. 39.

3 *Redemptor Hominis*, Art.

4 Address on Christian Unity in a Technological Age (Toronto, September 14, 1984), in *Origins* 14:16 (October 4, 1984),248, cited in *EJA*, Art. 94.

addressing the basic needs of the poor has to come before meeting the wants of the nonpoor, however good these wants may be in themselves.

The U.S. Bishops in their pastoral letter on the American economy came to the same conclusion.

> "The obligation to provide justice for all means that the poor have the single most urgent claim on the conscience of the nation."[5]

In the final sentences of their letter, the bishops eloquently expressed the call to social justice today as integral to living out the central gospel commandment of love.

> "We know that we are called to be members of a new covenant of love. We have to move from our devotion to independence, through an understanding of interdependence, to a commitment to human solidarity. That challenge must find its realization in the kind of community we build among us. Love implies concern for all—especially the poor—and a continued search for the social and economic structures that permit everyone to share in a community that is part of a redeemed creation."[6]

Another development in John Paul's most recent social teaching is the concept of *sinful social structures*. In 1987 he wrote of "structures of sin" which are "rooted in personal sin and thus always linked to the concrete acts of individuals who introduce these structures, consolidate them and make them difficult to remove. And thus they grow stronger, spread and become the source of other sins, and so influence people's behavior."[7] In 1984 he had spoken of "social sin" as rooted in personal sins "of those who cause or support evil or who exploit it; of those who are in a position to avoid, eliminate or at least limit certain social evils but who fail to do so out of laziness, fear or the conspiracy of silence, through secret complicity or indifference; of those who take refuge in the supposed impossibility of changing the world and also of those who sidestep the effort and sacrifice required, producing specious reasons of a higher order."[8] In other words, none of us can absolve her/himself of

5 *EJA*, Art. 86. See also, Articles 24, 89–94.

6 *EJA*, Art. 365.

7 John Paul II, *Sollicitudo Rei Socialis*, Art. 36.

8 John Paul II, Apostolic Exhortation *Reconciliatio et Paenitentia* (December 2, 1984),16. *Acta Apostolica Sedes*. 77 (1985), 217.

guilt for the existence of social injustice and the sinful structures which generate it.

Fred Kammer, S.J. has summarized this teaching by identifying sinful social structures as those which "destroy life, violate human dignity, facilitate selfishness and greed, perpetuate inequality, and fragment human community." He concludes: "As such they embody evil in the way sinful deeds do."[9]

The tradition of Catholic social teaching is rooted in the biblical teaching of both testaments, consistently calls for social justice as a hallmark of fidelity to God, has evolved over the centuries in response to the challenges of the times, and now calls us to give moral priority to transformation of societal structures which oppress the poor of our day.

Ecology in the Church's Teaching

The awareness of the ecological crisis of our time was not reflected until recently in the papal and episcopal social teachings. The 1986 letter of the U.S. bishops on the U.S. economy was criticized because it did not reflect new ecological understandings and the linkages between economic justice and environmental responsibility. Up to that time, papal social encyclicals had also neglected ecological concerns. But since 1986, both types of documents have begun to integrate these concerns into their teaching.

In his 1987 letter, *Sollicitudo Rei Socialis*, John Paul II identified as one of the positive signs in the contemporary world:

> "...a greater realization of the limits of available resources and of the need to respect the integrity and the cycles of nature and to take them into account when planning for development, rather than sacrificing them to certain demagogic ideas about the latter."[10]

Two years later, John Paul focused his message for the World Day of Peace on the ecological crisis, naming it a moral problem which calls for a new solidarity between rich and poor nations.[11]

9 Kammer, Fred. *Doing Faithjustice: An Introduction to Catholic Social Thought* (Mahwah, N.J.: Paulist Press, 1991), 9.

10 John Paul II, *Sollicitudo Rei Socialis*, Art. 26, *Origins*, 17:38 (March 3, 1988), 641–660.

11 John Paul II, Message for the 1990 World Day of Peace (January 1), released

In his 1991 letter, *Centesimus Annus*, the pope described the harmful effects of consumerism in the rich nations on persons, society and the environment. At the root of the destruction to the environment caused by overconsumption, says the pontiff, is a widespread anthropological error:

> "Man thinks that he can make arbitrary use of the earth, subjecting it without restraint to his will as though it did not have its own requisites and a prior God-given purpose, which man can indeed develop but must not betray. Instead of carrying out his role as a cooperator with God in the work of creation, man sets himself up in place of God and thus ends up provoking a rebellion on the part of nature, which is more tyrannized than governed by him."[12]

Despite its offensive sexist language, the text is significant as an illustration of development of church teaching on ecological concerns.

Later in 1991, the U.S. Bishops issued their pastoral letter "Renewing the Earth."[13] Calling the environmental crisis a moral challenge which is a call to conversion, the bishops reflected on the signs of the times, the biblical vision of God's good earth, Catholic social teaching and environmental ethics, theological and pastoral concerns, and concluded with a call to action. Acknowledging that Catholic social teaching does not have a complete environmental ethic, they described seven themes drawn from that tradition as integral to ecological responsibility, offering these as a basis for engagement and dialogue with science, the environmental movement, and other communities of faith and good will.

1. A Sacramental Universe. Knowing the universe as God's dwelling, people of faith are called to reverence for creation and to a stewardship which means living responsibly with and towards all creatures.

2. Respect for Life. Respect for life extends to all of creation; diversity reflects God's glory, and other creatures have value in themselves, not just as means to human fulfillment.

December 5, 1989 at the Vatican. Text was published in *The Church World*, (December 12, 1989), 6–7.

12 John Paul II, "Centesimus Annus," Art. 36, in *Origins*, 21:1 (May 16, 1991), 1–23.

13 *Origins*, 21:27 (December 12, 1991), 425–32.

3. The Planetary Common Good. The traditional teaching on responsibility for the common good first extended to the local community and the nation. In the 1960s, as global consciousness grew, John XXIII extended the concept to the whole human community.[14] Now we know in new ways that interdependence requires co-responsibility, and that preservation of the natural and human environment is integral to the common good.

4. A New Solidarity. John Paul II calls solidarity the moral response to the reality of interdependence, and describes it as the virtue which is "the firm and persevering determination to commit oneself to the common good..."[15] It requires sacrifices of our own self-interest for the good of others and of the earth we share. The ecological crisis reveals the need for new solidarity in relations between developing and highly industrialized nations to foster equitable sustainable development which avoids the environmentally destructive effects of what John Paul II has called the "superdevelopment" of the rich nations.[16]

5. Universal Purpose of Created Things. As we move toward environmentally sustainable economics, we are responsible to work for a just economic system in which the earth's bounty and the fruits of human work are equitably shared among all peoples. The rights of private ownership are limited by the needs of others. All are invited by God to share in the riches of creation—as neighbors—and all are responsible to care for the earth.

6. Option for the Poor. The poor suffer most directly from environmental degradation and have the least access to relief from their pain. "Nature will truly enjoy its second spring only when humanity has compassion for its own weakest members." The dignity of work and of workers must be preserved. We need to find solutions which "do not force us to choose between a decent environment and a decent life for workers."

7. Authentic Development. True development calls for moderation and even austerity in the use of material resources, invites development

14 See, for example John XXIII, *Mater et Magistra*, Art. 65,80 and *Pacem in Terris*, Art. 130ff in O'Brien David J. and Thomas A. Shannon (eds). *Renewing the Earth: Catholic Documents on Peace, Justice and Liberation* (Garden City, N.Y.: Image Books, 1977).

15 John Paul II, *Sollicitudo Rei Socialis* (1987), 38.

16 John Paul II, *Sollicitudo Rei Socialis*, Art. 28.

of alternative visions of the good society, requires affluent nations to reduce their overconsumption of natural resources, and encourages the search for agricultural and industrial technologies which respect nature.[17]

The statement of these principles reflects a basic anthropocentrism that is called into question by the new cosmology of thinkers like Thomas Berry. But, albeit belatedly, it demonstrates a growing integration of ecological understandings in the tradition of Catholic social teaching.

The call to action on behalf of justice and participation in the transformation of the world as a constitutive dimension of the church's mission[18] now clearly incorporates ecological responsibility as an element. To do justice one must act responsibly toward the environment. Social justice and ecological responsibility are inherently linked together.

Injustice and Ecological Devastation—In Current Structures

And injustice and ecological devastation are inherently linked together in the functioning of current structures. According to the 1999 Human Development Report published by the United Nations Development program, the gaps between the rich and the poor—within and among nations—are wider than they have been in half a century. The poorest 20% of the world's people experienced a decline in their share of global income from 2.3% 30 years ago to 1.0% today. Over the same period the share of the richest 20% of the world's people rose from 70% to 86%. The income of the richest is now 74 times that of the poorest, double the ratio of three decades ago.[19] Global economic and political structures are generating increasing inequality and deepening misery for the poorest of the world's people.

17 The seven themes are developed on pp. 428–430 of the *Origins* text of "Renewing the Earth."

18 Synod of Bishops. *Justice in the World* (1971) published in O'Brien and Shannon (eds), pp. 390–408. The cited passage is the last sentence of the Introduction found on 391.

19 United Nations Development Programme. *Human Development Report 1999* (New York: Oxford University Press, 1999), 2–3. and Barbara Crossette, "U.N. Survey Finds World Rich-Poor Gap Widening," *The New York Times* (July 15, 1996), A–3.

This kind of inequity is inexorably linked to environmental degradation. Sir Sidrath Ramphal of the World Conservation Union put the relationship succinctly: "At the root of all environmental problems lie the destitution of poor nations and the overconsumption of rich nations."[20]

Poverty causes environmentally destructive behavior. When population growth exceeds the rate of economic development, poor people are pushed onto fragile lands where there struggle for survival causes deforestation, exhaustion of soil and desertification, silting of rivers, destruction of habitats and extinction of species. Debt burdens force poor nations to focus their economies on earning foreign exchange through exports. This leads to sacrifice of food self-sufficiency, exhaustion of fisheries and forests, and use of harmful pesticides. Inappropriate transfers of technology cause waste of non-renewable resources, pollution of air and water, and destruction of age-old patterns of production beneficial to social harmony and environmental balance. Unless the problems of poverty are addressed, environmental destruction will continue as poor people seek to survive.

And overconsumption in the rich nations is an environmentally destructive behavior. With 20% of the world's population these nations generate 75% of the world's pollution, and consume 75% of the world's metals, 85% of its wood, and 70% of its energy. Replicating the patterns of the rich nations in the poor ones would require 10 times the amount of fossil fuel and 200 times as much mineral wealth as used by the rich nations now.[21] The richest 20% of the world's population have generated over the past century two-thirds of the greenhouse gases that threaten the world's climate. Their air conditioners, aerosol sprays and factories release almost 90% of the chlorofluorocarbons that destroy the earth's protective ozone layer. Their industries generate most of the world's hazardous chemical wastes, and their energy use each year releases as much as 75% of sulfur and nitrogen oxides that cause acid rain.[22] The

20 Quoted in Smith, Elaine, "Growth vs. Environment," *Business Week* (May 11, 1992), 69.

21 United Nations Development Programme. *Human Development Report 1994* (New York: Oxford University Press, 1994), 18.

22 Durning, Alan, "Asking How Much Is Enough," in Lester Brown et al. *State of the World 1991: A Worldwatch Institute Report on Progress Toward A Sustainable Society* (New York: W.W. Norton, 1991), 156.

people of the USA throw away each year enough paper and plastic plates and cups to feed all the people of the world a picnic six times a year. They discard enough aluminum cans to build 6,000 DC-10 airplanes. [23]

Current socio-economic-political-cultural structures which have generated the prosperity enjoyed by the few promote the injustice and ecologically damaging behaviors. But this is little understood by those of us who are the beneficiaries of these structures. Democratic socialist Michael Harrington called this our "cruel innocence" and believed that if we could come to understand, it could be the beginning of change.[24]

The structural causes of injustice and ecological devastation are inherent in the nature of the decision-making calculus of capitalist-market economies.

- The goal to which we are socialized is growth, private accumulation—always more.
- The primary value is individual freedom and self-interest. The basic premise is that if everyone takes care of her/himself, everyone will be taken care of. Individual self-interest leaves out the common good and the social goods which are necessary for the good society. Billions are spent on advertising things to satisfy personal wants, but public needs and the needs of those without purchasing power are not promoted at all.
- Nonquantifiable impacts are left out of the decision-making about what will be produced, how, and for whom. Ecological impacts such as soil erosion or the generation of nuclear waste are simply ignored in the cost-benefit analysis of corporations, since they do not have to pay the price for these impacts. The social impact on families and communities of downsizing or outsourcing or closing a factory is not factored in because the company does not have to pay for the effects of their decisions on those whose livelihood is destroyed.
- This partial costing process means that resource-using rather than resource-conserving technology is utilized, because the

23 *Ibid.*, 161.

24 Harrington, Michael. *The Vast Majority: Journey to the World's Poor* (New York: Simon and Schuster, 1977), 254.

full costs for present and future are not taken into account.

- Global corporations are free to roam the world in search of ever cheaper labor and permissive environmental regulations without any corresponding international law and structures to promote economic justice, protect workers and preserve the global environment on which we all depend for life. This means that globally we are now where we were in the industrial nations in the late 19th century—before environmentally protective government regulations and the struggles of labor unions and labor parties to gain safeguards for workers.
- Western models of industrialization are being copied in Asia, Africa and Latin America even though the resource endowments and the global environmental situation are far different today than when the West industrialized. Lack of research into alternative models, Western education of leaders, and transfers of technology through international agencies dominated by Westerners mean that resource-wasteful and environmentally destructive patterns of economic growth are being replicated.
- Economic growth has been the way out of poverty and still is essential to meet the needs of growing populations. But the content and structure of economic growth has to change. Production has to become less material-using and less energy-intensive, and distribution patterns have to become more equitable.[25] World consumption and income patterns have to be restructured.

This distorted calculus of the capitalist market economic order is so influential in generating injustice and ecologically damaging behavior because the economic dimension of life has such a pervasive influence on human life at this time.

For the poor, subsistence needs are survival needs, and so the struggle for economic resources is necessarily the critical focus of life. The daily search for food, water, wood and shelter consumes the waking hours.

Those of us among the world's privileged 20% are socialized to believe that we are what we have, that our value depends on how much

[25] See *Human Development Report 1994*, 20.

we earn, the kind of car we drive, the kind of neighborhood we live in. The society's valuation is communicated through advertising, through media-promoted status symbols. Alan Durning, author of *How Much Is Enough?*, observes:

> "When alternative measures of success are not available, the deep human need to be valued and respected by others is acted out through consumption. Buying things becomes both a proof of self-esteem and a means to social acceptance."[26]

John Kavanaugh, S.J. has eloquently and persuasively demonstrated in his book, "Following Christ in a Consumer Society," how the commodity culture dehumanizes both those who have an abundance of material things and those who are without basic necessities.[27]

Research by social psychologists has shown that Americans define happiness as having "more." Yet studies by the National Opinion Research Center reveal that no more Americans report that they are "very happy" than did in 1957 before the dramatic increase in consumption that has taken place over the past four decades.[28] Our frenetic quest for ever more is apparently not bringing us the happiness we seek, and at the same time it is contributing directly to the maintenance of social structures which generate both poverty and environmental degradation. And the stakes are very high. It is not an exaggeration to state, as does Durning:

> "In the end, the ability of the earth to support billions of human beings depends on whether we continue to equate consumption with fulfillment... In a fragile biosphere, the ultimate fate of humanity may depend on whether we can cultivate deeper sources of fulfillment, founded on a widespread ethic of limiting consumption and finding nonmaterial enrichment."[29]

In sum, the functioning of current economic and societal structures

26 Durning, "Asking How Much Is Enough," p. 162. His book expands on this article, *How Much Is Enough? The Consumer Society and the Future of the Earth* (New York: W.W. Norton, 1992).

27 Kavanaugh, John. *Following Christ in A Consumer Society: The Spirituality of Cultural Resistance* (Revised Edition) (Maryknoll, NY: Orbis Books, 1991).

28 Durning, 156.

29 Durning, 157, 165.

generate the forces fostering the poverty and overconsumption that generate environmental degradation. If we are to have a just and sustainable world, the people of the rich nations will have to face their responsibility for the systemic inequities that force the poor to abuse the earth in order to survive, and will have to come to understand that the earth cannot support everyone at their current level of consumption. And in the challenging words of Anil Agarwal and Sunita Narain of the New Delhi Center for Science and Environment, the poor nations will have to work to eliminate their poverty and to "renounce the allure of Northern lifestyles and stand for integration in modern living of the ecological prudence of traditional lifestyles."[30]

ALTERNATIVE VISIONS OF THE GOOD SOCIETY

Christians, sharing the prophetic mission of Jesus, are called to announce the possibility of an alternative to the sinful structures we denounce. Over the past two decades, as the present effects and future implications of the just-described realities became more and more apparent, a growing group of thinkers, educators and activists from various disciplines have begun to develop alternative visions of just, ecologically and socially sustainable societies, and the paths to attain them. And through the interdisciplinary dialogues among concerned global citizens and the innovative projects on various levels around the world the shape of what could be is emerging.

Hazel Henderson, one of the nation's original environmentalists, widely recognized as one of the most innovative thinkers of our time, has been writing and speaking since the mid 1960s about the distorted calculus of economic-decision making which leaves out of its reckoning of costs and benefits all the contributions of non-paid labor and the world of nature, and all of the impacts of laws, customs, culture which operate in the nonmonetized social system.[31]

Internationally recognized political economist, Marilyn Waring of

30 Agarwal, Anil and Sunita Narain, "View from the South: We can No Longer Subsidize the North," *Development Forum*, 20:3 (May–June, 1992), 15.

31 See for example: *The Politics of the Solar Age* (New York: Anchor/Doubleday, 1991); *Creating Alternative Futures: The End of Economics* (New York: Berkley Publishing Corporation, 1978); *Building a Win-Win World: Life Beyond Global Economic Warfare* (San Francisco: Berrett-Koehler, 1996).

New Zealand, in her 1988 book, *If Women Counted*, synthesizes the research on the invisibility of women's labor and the lack of an imputed value for the environment, and shows how and with what effect these invisibilities have been institutionalized in national and international accounting systems and public policies.[32]

A group dubbed "ecological economists" have been challenging their colleagues to recognize the limitations of analysis which undervalues environmental resources and impacts. Primary among them is Herman J. Daly, who has been probing these issues for over two decades. His early writings include *Steady-State Economics: The Economics of Biophysical Equilibrium and Moral Growth*, which presents an integrated model of an economy characterized by redefinition of the goal of economic growth as increased quality of life, maximum and minimum limits on income, and transferable depletion quotas to limit non-renewable resource use and therefore pollution.[33] With theologian John B. Cobb, Jr., Daly published in 1989 *For the Common Good: Redirecting the Economy Toward Community, the Environment and a Sustainable Future*, a searching critique of how our growth-oriented, industrial economies have caused environmental disaster. The authors present a new paradigm for economics, social ethics and public policy in their economic model of a world order based on self-reliant communities. Economies would be driven by service to community; trade would be balanced with priority given to local self-sufficiency in necessities; regionalization would be fostered to prevent outflows of capital. Theistic faith is seen as a way to integrate both the goals of a humane society and ecological sustainability.[34]

British economist James Robertson has developed a model of a "SHE future"—sane, humane and ecological—in which persons are in harmony within themselves, with one another and with nature. He envisions a multi-level one-world economy consisting of autonomous and

32 Waring, Marilyn. *If Women Counted: A New Feminist Economics* (New York: HarperCollins, 1988).

33 Daly, Herman J. *Steady-State Economics: The Economics of Biophysical Equilibrium and Moral Growth* (San Francisco: W.H. Freeman, 1977), Chapter 3.

34 Daly, Herman J. and John B. Cobb, Jr. *For the Common Good: Redirecting the Economy Toward Community, the Environment, and a Sustainable Future* (Boston: Beacon Press, 1989).

self-reliant but interdependent parts, with institutions on a humane scale.[35]

David Korten spent thirty years in Asia, Africa and Latin America as a writer, teacher and management consultant. This experience convinced him that radical rethinking of development—linking economy, ecology and spirituality—is necessary. Korten founded The People-Centered Development Forum, a global networking organization to advance debate and education on alternative development strategies. Korten proposes a people-centered development model based on growth in quality of life rather than material consumption, responding to the needs of all rather than the wants of the monied, fostering community cooperation rather than individual competition, striving for local diversification and self-reliance in basics rather than for specialization for international markets, and encouraging financial and environmental conservation and saving rather than borrowing and debt.[36]

While the thinkers have been exploring the issues in depth, others have been working at local levels to create from the bottom up just, ecologically and socially sustainable societies.

The Irish Columban missionary Sean McDonogh and his colleagues worked with the T'boli, an indigenous group in the Philippines whose livelihood was being threatened by environmental degradation. With the people they discovered how to divide the land into plots so that with crop rotation, animal husbandry, recycling of by-products they achieved sustainable ecological balance in the tribe's habitat. Also in the Philippines, the Green Forum composed of cooperatives and non-governmental organizations (NGOs) working for environmental and social change issued a detailed plan for an alternative development economics—community centered capitalism.[37] Through linkage with U.S.

[35] Robertson, James. *The Sane Alternative: A Choice of Futures* (St. Paul, MN: River Basin Publishing Company, 1978)and *Future Wealth: A New Economics for the 21st Century* (London: Cassell Publishers, 1990).

[36] Korten, David. *Rethinking Development: Economy, Ecology & Spirituality*. Video and Study Guide (New York: Codel, Inc., 1993); "Sustainable Development: A Review Article" World Policy Journal, 9:1(Winter, 1991–92), 158–190; *When Corporations Rule the World* (San Francisco: Berrett-Koehler, 1995).

based organizations the Forum has already succeeded in getting $15 million of U.S. foreign aid money earmarked for environmentally responsible projects sponsored by the member organizations in the Philippines.

Appropriate technology for economically and ecologically sustainable projects in the developing world is developed by CERETECH, a company whose staff is experienced in sustainable development in the Middle East and Africa. The company with offices in Chicago, Zagreb and Berkshire in the United Kingdom specializes in international rural development and fosters manufacture by local artisans of the lathes, rice strippers, grinders, spinners, had drills which it has developed. The tools can be either hand or foot powered, are of simple and sturdy construction from materials readily available in most developing countries.

The best of the sustainable human development approach is embodied in Bangladesh's Grameen Bank, a commercial bank dedicated to alleviating rural poverty through microenterprise lending in a process which builds solidarity as well as self-reliance. Its repayment rate of 98% exceeds that of the world's largest banks; it reaches more than ten million poor, rural households in Bangladesh with small loans (equivalent to $60-70 on average); it has enabled measurable improvement in the income and assets of its borrower-members, 91% of whom are women.[38] The model has been replicated in Latin America by ACCION International, which is now also implementing it in the inner cities of San Antonio, Los Angeles and New York. Microenterprise lending usually operates within the informal sector where millions of the world's poorest urban residents eke out a living. Though some kinds of activity, such as charcoal production, can be environmentally harmful, much of informal sector activity is environmentally healthy—recycling materials discarded by households and businesses of the formal sector.

Cooperative ventures among women are growing in Latin America and Africa, enabling women to share skills and techniques in weaving, soap making, vegetable gardening, and energy utilization.[39] India's 40,000 member Self-Employed Women's Association has become a development

37 Philippine Green Forum. *An Alternative Development Economics: Economic White Paper* (Manila: Green Forum Philippines, 1991).

38 Holcombe, Susan. *Managing to Empower: The Grameen Bank's Experience of Poverty Alleviation* (London and New Jersey: Zed Books, 1995).

39 Benson, Nancy, "African Women Cooperate to Create Development

model for low-income women around the world.

In the aftermath of the wars in Nicaragua and El Salvador, farmers returning to their land are learning new methods of farming without chemical fertilizers and pesticides.[40] For example, the SHARE Foundation is sponsoring a training program in organic farming in which 15 communities and cooperatives in Tecoluca, El Salvador are participating. Other groups are working to link producers in Central America with buyers of organic products in North America and Europe. Innovative partnerships like these, funded in large measure by socially conscious U.S. donors and investors, are making environmentally sound sustainable development a reality for the people of Central America.

These few examples demonstrate that seeds of change are being sown in many directions and places. New thinking is being sparked by iconoclasts. New economic models are being proposed by ecological economists and by development management experts. And in local communities around the world, microenterprise lending, women's cooperatives, and organic farming are already operating out of alternative visions of development, seeding a future in which justice and environmental and social sustainability can be realities. There are alternatives to the current structures and they are rising from the bottom up.

Implications for Action

If these seeds are to come to fruition, a process of social transformation is required. This calls for many different kinds of action by people committed to a vision of a just, humane and sustainable world in which all persons will be able to live in dignity and in harmony with all of creation. Many people thinking globally and acting locally with a new intentionality informed by this vision in one or more of the necessary action roles can effect social transformation.

A simple paradigm of eight roles summarizes all of the types of action that are necessary and sufficient for systemic change.

• Study and teaching. Some must do the research and education

---

Alternatives," *Listen Real Loud: News of Women's Liberation Worldwide, 12:1* (1993), 26–28

40 See for example the special issue on sustainable agriculture of *Grassroots Development: Journal of the Inter-American Foundation*, 19:1 (1995).

around the issues, the moral principles, and the alternative approaches—and communicate widely.

- Building the new value system. Parents, pastors, teachers, artists, musicians, actors will need to help make attractive the values that will counter the socially and environmentally destructive commodity culture.
- Choosing lifestyles consistent with the ethic of justice and ecological responsibility. More of us have to simplify our lifestyles, learning to eat, clothe ourselves, transport ourselves, and recreate in less resource-using ways—so that others may simply live.
- Creating and supporting alternatives. Some have to create alternatives that embody the new vision; others must support, publicize and nurture the new ventures.
- Transforming existing institutions from within. Working within existing institutions—family, church, business, education, health care—the values of justice and ecological responsibility have to be implemented, subtly reshaping the institutions and the culture that supports them.
- Strategizing and organizing politically. As citizens, people have to take seriously the issues of justice and ecology, participating in organizations which develop strategies for change
- Standing in opposition to everything that goes in the wrong direction. Saying "no" to current abuses can take many forms: participating in boycotts, writing letters to the editors, calling in on radio talk shows, marching, civil disobedience. But in the face of the structures of sin doing nothing is not an option.
- Living out of an integrated, kin(g)dom-centered spirituality. All people of faith are called to work to overcome the dualisms that have allowed the current abuses, living out of trust in the enabling power of God who calls us to be agents of transformation and is ever making all things new.

No one can say there is nothing s/he can do to foster social transformation, and thinking of it in terms of these action roles gives a glimpse of the transforming power of a group of committed people acting according to their own gifts in their own spheres of influence—family, church, business, government, professions, trades—with a new intentionality focused on social justice and ecological responsibility.

Clearly, the challenges of justice and ecological responsibility call us to personal conversion as well as social transformation. Today there is much talk of holistic spirituality. Perhaps there is need to resist the temptation to easy faddism which wants to focus on comfortable, consoling thoughts and images to the neglect of the realities. A true creation spirituality in today's world has to bring us face to face with poverty and environmental degradation. We cannot divorce ourselves from the quest for social justice and ecological co-responsibility.

Continuing effort is needed to develop a socially and ecologically responsible conscience which will guide thought, choice and action. Living in an individualistic, materialistic, wasteful culture challenges us to be countercultural in order to be faithful. Slogans such as "Recycling is good; reusing is better; reducing is best," can be added to our memory bank. Deliberately cultivating a global consciousness, and putting human faces on the statistics of poverty can be a powerful means to the needed re-formation of conscience.

To sustain efforts to learn to live with enough, it will be important to reflect on how tis will bring new fulfillment and joy. A statement attributed to Thoreau notes that a person is rich in proportion to the things s/he can afford to let alone. And Alan Durning's words on this topic are rich in meaning:

> In the final sense, living with enough rather than excess offers a return to what is, culturally speaking, the human home: to the ancient order of family, community, good work and good life; to a reverence for the excellence of skilled handiwork; to a true materialism that does not just care *about* things, but cares *for* them; to communities worth spending a lifetime in.[41]

Justice and ecological responsibility are linked in the moral principles of Catholic social teaching and in the crises of poverty and environmental degradation in our world today. As Christians our call is to understand, to respond, and to commit ourselves to the personal conversion and social transformation required to bring about a world in which all persons can live in dignity and harmony with all of creation.

41 Durning, 169.

Christianity *in an* Emergent Universe

Thomas Berry, CP

Some twenty years ago I attended a conference of religious educators sponsored by the archdiocese of Buffalo, New York. Over two thousand teachers at the elementary and high-school levels were there. I spoke on the need to teach the sacred dimension of the universe. Afterwards I spent a long time looking over the literature for sale there by over thirty different book companies. There I found books on belief, on the bible, the commandments, prayer, the Eucharist; on Jesus, the Blessed Virgin, the saints; on social issues, on religious vocations, on marriage; on a multitude of subjects.

What I did not find was anything on the universe or the earth, on creation, except for assertion of God as Creator. Nothing on evolution, or on the natural world. I found no presentation of the universe as the manifestation of the Divine. Supposedly the universe belonged to a different realm of things, a realm that was not of any special concern for religious education.

Already at the time it was evident that the earth was being devastated, that there was no future for the human community if the integral functioning of the planet were not protected. Yet this was considered irrelevant to the religious instruction of the young, even though they would soon find themselves living amid the ruins of the industrial world and amid the ruins of the natural world itself. To me this was a cause and a consequence of the failure to understand that religion itself is born out of the wonder and magnificence and fearsomeness of the natural world.

Combined with this neglect in taking the universe or the present situation of the planet earth seriously throughout elementary and secondary education, this situation is continued through the entire range of Catholic educational establishments; by the seminaries, colleges and universities. Also by the religious orders. The Jesuits, the most powerful educational establishment throughout the entire human community, with some hundreds of educational establishments in every region of the globe, is not distinguishing itself in its ecological concerns.

One of our other most prominent Catholic universities recently for their summer session was advertising courses in medieval thought, in the bible, in the usual Catholic theological concerns, with no indication of any interest in the manifestation of the divine in the world about us as undergoing at the present time a terrifying and in many ways an irreversible devastation. Whatever interest there is in cosmology or the natural world is in terms of medieval thought, not in any immediate experience or in terms of contemporary empirical inquiry into the structure and functioning of the universe. The universe is given over to the scientists or to the economists or to the medieval theologians.

Meanwhile the planet continues to deteriorate to such an extent that biologists such as E. O. Wilson and Peter Raven are comparing the present order of magnitude of disruption of the life systems of the planet to the catastrophic destruction that occurred at the end of the Mesozoic period some 65 million years ago. In this larger perspective, as regards the future of the human community, our most significant need is to know where we are in the larger story of the universe and how to establish a viable human role within this context. Until recent times this was less important since the universe was thought of as an abiding sequence of seasonal transformations without beginning or ending. Discovering our role at that time was less critical, since we had not the power to interrupt the process.

The distinctive feature of the bible was to assert that the universe had a beginning in time, that humans were undergoing an historical process in arriving at the Day of Lord at the Messianic Age. The King-dom of God was coming into being. The Mystical Body of Christ was in formation. The Millennial period would one day arrive. The manifestation of the divine was more in historical events such as the Exodus or the revelations given to the prophets or the Incarnation of the Savior. All of this would happen in a human historical order, but in a non-historical universe. Once the universe came into being it had no story, except indirectly through the human story.

Human identity did not need a historical sense in the same manner as we in the 20th century need a sense of our place in the historical sequence, for now we think of ourselves as existing within a universe that had a beginning in the distant past and has become what it is through a long sequence of irreversible transformations moving in the larger arc of their development from lesser to great complexity, from lesser to greater consciousness, from lesser to greater freedom. The human story is integral with the universe story.

This sequence of irreversible transformations is now the central fact of our existence. We and everything around us came into being within this sequence of transformations. Our personal story is intelligible only through this larger story since everything that has happened in this more comprehensive story has been needed for us to be as we are. To know ourselves in any comprehensive manner is to know the story of the universe. The universe is the larger self of each individual reality in the universe. The universe story and our human story are so integral with each other that none of these can be fully understood without the others. This universe that we now live in is infinitely more exciting and more meaningful than simply an abiding seasonal universe, a universe that was essentially complete in its structure from earlier times when all things were thought of as having come into being essentially as they are at present.

The historical changes that have occurred are traditionally thought of as brought about by moral or social causes rather than by evolutionary processes inherent to the natural world. Discovery of historical-developmental time through empirical inquiry into the structure and functioning of the universe has altered our perception of the universe and of ourselves in a manner that we have not yet been able to assimilate into our Christian mode of thinking.

Our sense of the sacred is so integral with the earlier understanding of a universe going through an ever-renewing seasonal sequence that we find it difficult to accept the sacred dimension of a universe coming into being through a sequence of irreversible transformation episodes. Since it has evolved from within its own dynamism, this universe is considered to be alienating from any sense of having its origin and its controlling force within a divine reality, or as being the manifestation of divine reality. Such indeed is the manner in which this new explanation of the universe has been presented to us. In this context both the universe and the human are seen as lacking in that spiritual form that they possessed in the earlier modes of thinking of the universe and of the human presence in the universe.

In reality it should be quite simple to realize that the capacity of the universe for self-evolution in the phenomenal order can be understood as a capacity given by a transphenomenal source. A different and even a more sublime sense of creation can be established within this new context. It is simply to take seriously the teaching of Saint Thomas concerning the reality of phenomenal causality. This distinction of phenomenal and trans-phenomenal causes implies also that these are two aspects of a single energy. The divine creates a world that creates itself. As indicated by Saint Thomas, the divine acts in every being that acts. The divine brings us into being, sustains us in being and moves us toward our proper destiny, all within this process of self-creativity whereby the universe and each being in the universe creates, sustains, and moves itself toward its proper destiny.

Since this new understanding of the universe as evolutionary emergence has arisen within the context of the Christian humanist western tradition, we might think that Christians would have a certain inherent ability to appreciate and to accept this new sense of the universe within the context of their own thinking. Yet this has not happened. One reason is that this new understanding of the universe has brought with it a certain control over the Earth. The capacity of humans to obtain their food, to shelter and clothe themselves, their capacity for travel over the land and the sea and even through the air; all of this has given to the human a certain arrogance, a sense of independence from any trans-phenomenal power.

Thus a strange thing has happened to our western scientific world. By losing our awareness of the divine presence in the universe we lost our reverence for the universe and for the planet Earth. We might have thought that the universe in its independence from any divine reality would have become doubly precious since it would present a phenomenal world that was self-originating and self-sustaining. The universe, we might have thought, would itself take on divine qualities.

In reality, our western society absorbed into the human all those qualities formerly associated with the divine. Then a discontinuity was established between the human and the non-human worlds. All inherent values and rights were attributed to the human. The non-human had no inherent value and no inherent rights. Its only value was use by the human. A profound misconception concerning the Earth took place: the Earth was seen as a limitless resource for exploitation by humans. This became the functioning attitude of western industrial society toward the planet. Science became subservient to economics. Science-based technologies enabled the human to exploit the other components of the

Earth community in merciless fashion, with consequences that only now begin to be understood in the full terror of their implications.

We seem determined to take the planet to pieces in the expectation that we would be able to put it back together in a manner that will be less oppressive than the natural world. Yet we suddenly find that we are creating, not wonderworld but wasteworld, a world of ever-accumulating junk, rather than the magical world that we thought we were bringing into being. We now find that the natural life systems cannot absorb the devastation we are causing or the poisons that we are pouring into the air, the water and the soil.

A renewal of the vital forces of the planet, indeed the survival of the planet in any integral expression of itself, now requires that we move into a new creative period that might be designated as the Ecozoic Era, the period when humans would be present to the Earth in a mutually-enhancing manner. Nothing now before the human community is of such importance as this transition from the terminal Cenozoic into the emerging Ecozoic Era. The 21th century can be considered the century of the great Transition.

The greatest division in the human community and all human institutions and professions is now the division between those whose loyalties and values are committed to a continued exploitation of the planet in the late Cenozoic and those committed to creating a new integral mode of life on the planet designated as the emerging Ecozoic. This division will in the future give to individuals their public identity such as was formerly given by the terms "liberal" and "conservative" based mainly of social orientation. So too in the professions; in medicine, law, education and religion. Those within the various professions will be identified according to how they envisage themselves as functioning as a subsystem of the Earth system, or seek to dominate the Earth system.

To each of the professions might be said what can be said of education. Ecology is not a course. It is not a program. It is the foundation of all programs, all programs, all professions; because ecology is a functional cosmology and any understanding within the cosmos depends ultimately on an understanding of the cosmos. So ecology is not a part of law or medicine or economics. All these are parts or functions of ecology. So too in religion. Ecology is not primarily a subject of religion. We learn our religion from the universe. Religion is derivative from the wonder and majesty and fearsomeness of the universe or the cosmos.

The universe is the primary sacred community. As humans we join a preexisting sacred community. We do not originate the sacred community. The sacred community originated with the universe itself.

Christians should begin thinking in terms of an Ecozoic Christianity. That is a Christianity based on the great principle that the universe is coherent with itself throughout its vast extent in space and throughout its long sequence of transformations in time. The universe is the only self-referent mode of being in the phenomenal order, the only text without a context. Every other being is universe referent, a text within the universe context. So too the human. We too are a derivative mode of being. We are a subsystem of the earth system. Ultimately we are less beings in the universe than a mode of being of the universe.

In this setting it can easily be seen that the universe is the primary self manifestation of the divine, our primary revelatory experience. Verbal revelation can never be primary revelation. As soon as divine communication enters into human language it takes upon itself the peculiarities of the language, of the social structures, of the historical moment. Understanding the message in this context is too difficult. Our primary experience is more immediate with our daily experience, as anyone could testify who has reflected on their experience of the stars in the heavens or the wonders of earth, on the dawn or sunset, on the mountains in their grandeur or the oceans in their vastness, or the song of the mockingbird or the flight of the raptors. The impress of the divine is everywhere. In the air we breathe, in the food we eat, in the evening wind.

Saint Thomas tells us in the Second Book of his *Summa Contra Gentiles*, chapter 45, that "The order of the universe is the ultimate and noblest perfection in things." From this quotation and from a limitless number of other available quotations from biblical and Christian sources we can immediately appreciate the importance of the universe in Christian thinking. In mentioning that the "Order of the Universe" is the "ultimate and noblest perfection in things" we can appreciate how significant the universe is in his thinking; for he does not say the human or Christianity or the church, he says simply that the Order of the Universe is the final referent in the phenomenal world. From this it is clear that the human is for the perfection of the universe. The universe is not here primarily to serve the human. So too the Incarnation and Redemption are primarily for the perfection of the universe, not primarily for the human—even though these may be immediately for the human.

Of course to Saint Thomas, The Order of the Universe was understood quite differently from our present way of understanding the universe. For him the Order of Things was the Ptolemaic universe of spheres enclosed in each other with the Earth at the center and with humans the principal beings on Earth, a process that had been going on from eternity as far as we could know from our own intellectual thinking processes. His view was that we could never prove from any reasoning process that the world had a beginning in time. He had no idea of such enormous periods of galactic and geological and biological time or the sequence of transformational episodes as we now think about when we think of the universe. Nor could he have known that the human has emerged out of such a sequence. Undoubtedly he would have appreciated the wonder of such an emergent universe if our present knowledge had existed in his time.

This developmental universe should be understood as the context for any vital Christian development in the future. We have vast resources in our spiritual, intellectual and social traditions for dealing with this issue. The single most important resource, to my mind, is in the understanding of the universe and the place of the human in the universe given to us in the writings of Teilhard de Chardin. He is the first Christian theologian to understand and accept in a profound manner the story of the universe as coming into being through a sequence of evolutionary transformations. He understood fully that the universe from its beginning had a psychic-spiritual as well as a physical-material dimension; that the human in its every aspect is integral with the universe; that greater emphasis should be given to our understanding of creation in our religious-theological teaching.

Within this context Teilhard brought about what might be considered the most profound transformation of Christian Theology that has taken place since Saint Paul, whose teaching is considered as the foundation of Christian theological thinking. Teilhard's teaching can be considered the original statement of the Ecozoic phase of Christian thought.

One other important aspect of Christian thinking in the Ecological period is the sense of Cosmological Moments of Grace, for without cosmological moments of grace there could not be historical moments of grace or religious moments of grace. These moments of grace can be identified as the great transition moments in the story of the universe. For indeed this universe story, as the Epic of our times, can also be considered as the Christian Epic.

The earliest emergent moment can be considered such a moment of grace, also the moment when the galaxies were formed. A very special moment of grace might be indicated as the moment of the Supernova collapse and explosion into space of that special first-generation star that formed the materials out of which the solar system, the sun, the planets and especially the planet Earth were shaped. Prior to this supernova explosion the planet Earth, life and self-reflective consciousness could not come into being, for only a few of the lighter elements existed. It took this sacrificial deed for the ninety-some elements to take shape that were needed for such a planet as the Earth, with all those conditions required for life and the human mode of consciousness.

Other Moments of Grace could be enumerated such as the discovery of sexuality without which the present development of the living world could not have taken place. Also there is that stunning moment when photosynthesis was discovered, the way by which the energy of sunlight together with material from earth was used to create organic matter. So too the coming of the flowers some hundred million years ago. The coming of the flowers was a dramatic moment. Suddenly flowers in all their diversity of expression took over the planet from the immense green vegetation to produce a brilliantly colored planet. It was also a needed development of the concentrated protein in the covering found in seeds and nuts and berries and all those fruits and vegetables that were needed by the larger mammals whose brain development could hardly have taken place and whose energy requirements could hardly have been supplied if they had to be nourished by eating a diet of greenery such as covered the continents until this time.

Only when some of these moments are ritualized will they begin to enter into our religious consciousness. Then we can truly begin to think of the universe story as our Sacred Story. Only when this takes place, only when we begin to experience this new telling of the story of the universe as sacred narrative, will we experience the creative dynamic of this new understanding of creation. Once we have done this, then Christians can take their proper role in bringing about this new geo-biological-cultural era in the Earth Story, the era when humans will be present to the earth in a mutually-enhancing manner, the era when humans will truly join in that greatest of all liturgical celebrations, the celebration that is the universe itself.

We can do much to limit the present devastation of the planet out of our understanding of the universe given to us through the Book of Genesis but in my view, it will be inadequate and eventually ineffective in its larger dimensions, for it will not enable the Christian effort at creating the Ecozoic Era to join effectively with the wider effort of humans who work within the context that I have indicated.

Afterword

Margaret R. Brennan, IHM

In *Letters to A Young Poet,* Maria Rainer Rilke counsels a young nineteen year-old student who is asking for advice about becoming a writer himself. I suspect that all of us have read and pondered the following lines at one time or another and have discovered that they have had meaning in our own lives even if we are much more than nineteen years old!

> "...have patience with everything unresolved in your heart and try to love the questions themselves as if they were locked rooms or books written in a very foreign language. Don't search for the answers, which could not be given to you now, because you would not be able to live them. And the point is, to live everything. Live the questions now. Perhaps then. Someday far in the future, you will gradually, without even noticing it, live your way into the answers."

No longer young, not an aspiring poet, but hopefully ever a student, I think I can say that during most of my 75 years—and certainly during the 54 that I have been a religious—I have been at least willing to embrace the questions of faith, belief, church and mission that I have encountered, while acknowledging at the same time that I have not always done so gracefully.

In this volume of essays, colleagues, companions, and former students, many of whom have become friends and mentors, address the breadth and depth of many of the questions that have touched my life personally and continue to challenge my existence as an American, a Roman Catholic, a woman, an educator, a member of a religious congregation. Reading these essays has been a rich and rewarding experience. In them I have walked again on the pathways of my search for the living God within the church and within the world. In them I have seen once more, and sometimes in ways that are quite new, how inevitably, intricately, and intimately that journey is related to the emergent universe in which we live.

In my fifty-four years as a religious, different questions have arisen reflective of two distinct periods in the life of the church and now in the early years of a third. From 1945–1960 I lived in the pre-Vatican church where answers were clear and structures unchanging for the most part. Emerging questions were held tentatively and their public expression not favored. The years from 1960 until late in the seventies were ones when the spirit of "aggiornamento" allowed the old questions to surface freely and new ones to enter through the windows and doors that the church opened onto the modern world from which it had closed itself off in so many ways. I think, moreover, that many of us cherished a hope that the new questions would lead to answers and allow for a renewed church that would move creatively forward in the spirit of the Vatican documents—open to the world and welcoming of a diversity that was at the same time rooted and secure in its unchanging tradition.

The late seventies, eighties and nineties have made us aware of the demise of the myth of progress that characterized the modern world as well as the attendant evils that followed in its wake. The global context in which the church lives and moves today is one in which the centrality of western civilization has been challenged. The emergence of a polycentric world has brought a new summons to shape a spiritual and social vision that honors "the other." For want of a better term, the name postmodern is being used, among other things, as referent for the call to honor radical plurality and to be sensitive to what the differences in class, gender, race and culture mean in the articulation of our Catholic Christian Tradition.

Our time is not one in which answers are readily available, nor should they be. Rather, in the thought of Rilke, it is more of a summons "to live the questions," with the hope that one day, without noticing it, we may also "live our way into the answers." It seems to me then, that the "heavy blessing" of living and ministering in the church today is the challenge to live with ambiguity, to have the wisdom to know how to avoid an "either/or" in dealing with questions of continuity and discontinuity with regard to the treasures of the Christian Tradition. Living with ambiguity and the avoidance of "either/or" positions is to live with different levels of meaning, to explore their richness and complexity.

"...Every scribe who becomes a disciple of the kingdom of heaven is like a householder who brings out from his storeroom things both new and old."

These words of Jesus found in the gospel of Matthew (Mt.13:51–53) are descriptive of these friends and colleagues who have reached into their knowledge and experience to share some insights about life, ministry, and meaning in the Church today. I want to thank them for sharing their expertise, for raising some new question, and for pointing toward ways of living into the answers.

Chronological Bibliography *of the* Works *of* Margaret R. Brennan, IHM

Elaine M. Biollo, SC, Compiler

ARTICLES

1972

Not What Paul Said—But What He Didn't Say." *National Catholic Reporter* 9 (December 8, 1972): 7–8.

"Music For Our Times." *Sisters Formation Bulletin* 17 (1972): 5–16.

1973

"The Hard Questions of the Beatitudes." *Origins* (1973): 161–164.

1974

"Expanding Ministries for Religious Women Apostolic Spirituality." *Soundings* (1974): 28–30.

"Women in the Family of God." *Catholic Mind* 72 (1974): 38–46.

1975

"Women's Liberation/Men's Liberation." *Origins* 5 (1975): 98–105. Also published as "Disturbing the Perceptual Patterns: A Reflection on the Liberation of Men." Washington: Leadership Conference of Women Religious of the U.S.A., 1975.

"Standing in Experience: A Reflection on the Status of Women in the Church." In *Proceedings of the Canon Law Society of America* 37th Annual Convention (1975): 12–25; Also published in *Catholic Mind* 74 (1976): 19–32; and partially in *Origins* 5 (1975): 259.

1976

"Why We Need Women Priests: An Interview." *U.S. Catholic* (1976): 22–27.

Wintz J. "Women: The Church's Newest Spiritual Guides: Seven Women Talk About Their New Roles." *St. Anthony Messenger* (1976): 22–27.

"Religious Life in the Light of Future Ministry." Chap. in *Facets of the Future: Religious Life in U.S.A.* Ruth McGoldrick, SP and Cassian J. Yuhaus, CP., eds. Washington: Cara Press.

1977

"Was It Not Ordained That Christ Should Suffer and So Enter into His Glory?" *New Catholic World* (January/February, 1977): 220–221.

1978

"Renewal of Apostolic Religious & Renewal of Contemplative Religious." *New Catholic Encyclopedia*. Washington: McGraw-Hill.

"Was It Not Ordained That Christ Should Suffer and So Enter into His Glory?" *New Catholic World* (January/February 1977)

"Religious." *New Catholic Encyclopedia*, 1978.

"What Kind of Ecclesiology Will We Need In the Future?" *New Women, New Church, New Priestly Ministry* in Proceedings of the 2nd Conference on the Ordination of Roman Catholic Women, 96–100. Rochester, New York: Kirk-wood Press for the Conference on the Ordination of Roman Catholic.

1979

"The Role of Religious Institution in Evangelization." *Hospital Progress* 60: 63–67.

1980

"Women and Men in Church Office." *Concilium* 134, *Women in a Men's Church* 102–109. Seabury Press: New York.

1981

"Dialogue on Women in the Church: Interm Report." *Origins* 2 Vol. 11 No. 6 (1981):83–91.

1983

"Sing a New Song Unto the Lord: Spirituality and Social Justice." *Education for Peace and Justice* ed. Padraic O'Hare, 211–222. San Francisco: Harper and Row, 1981.

1984

"The Experience of the Essential Elements in the Lives of U.S. Experience." *Religious Life in the U.S. Church: The New Dialogue* eds. Robert Daly SJ; Michael Buckley SJ; Mary Ann Donovan SC; Clare Fitzgerald SSND; John Padberg SJ, 134–139. New York: Paulist Press, 1984.

1985

"Women and Theology: Singing of God in an Alien Land." *The Way Supplement* 53 (1985): 93–105.

1986

"Being Fully in the World." *The Way Supplement* 56 (1986): 54–64.

"Commentary on the Decree on the Appropriate Renewal of the Religious Life: *Perfectae Caritatis*." *The Church Renewed: The Documents of Vatican II Reconsidered*, ed. George P. Schner, SJ, 63–72. Washington: University of America Press, 1986.

1987

"Patriarchy: The Root of Alienation from the Earth?" *Thomas Berry and the New Cosmology* eds. Anne Lonergan and Caroline Richards, 57–63. Mystic CT: Twenty-Third Publications, 1987.

1988

"In the End the Lion is God." *The Way* 28 (January, 1988): 5–19.

"Authentic Self-Realization." A Catholic Bill of Rights eds. Leonard Swidler and Herbert O'Brien. Kansas City, MO: Sheed and Ward, 1988.

"John Paul has missed the meaning: Response to Pope John Paul's *Meditations on The Dignity of Women*." *Catholic New Times* (October 23, 1988).

"Feminism - Vision of a New World." *Toronto School of Theology Times* 21 (1988): 26–29.

Odell Catherine M. "Two Tell Seminar What's Right With the Church." *Our Sunday Visitor* 77 (May 8, 1988): 41–43.

1989

"The Blurred Blessing of Matthew Fox." *Compass* 7 No. 4 (January 1989): 31–33.

1990

"Teresa of Avila: Undaunted Daughter of Desire." Chap. in *Spiritualities of the Heart* ed. Annice Callahan, RSCJ, 114–129. New York: Paulist Press, 1990.

1991

"Why Women of Faith are Changing Their Metaphors." *Catholic New Times* (January 6, 1991): 11.

1992

"It's Like Looking At God." *The Way* 32 (October, 1992): 281–291.

1994

"Religious Life in a 'New World Order'." *Catholic New Times Supplement* (February 20, 1994): 1–8.

"Being a Religious Today: An Exchange of Notes on Contemporary Religious Life. A Round Table Discussion with Bernard Audigier, Margaret Brennan, Jean Doutre and Christine Leyser." *Compass* (May/June, 1994): 13–18.

"The Body and Our Likeness to God." *Compass* (1994): 18–20.

"A White Light Still and Moving: Religious Life at the Crossroads of the Future." In *Religious Life: The Challenge of Tomorrow* ed. Cashan Yuhaus. New York: Paulist Press, 1994.

1995

"Transforming Life: The Charism of Religious Life." In *Living in the Meantime: Concerning the Transformation of Religious Life* ed. Paul J. Philibert, 153–163. New York: Paulist Press, 1995. Originally appeared in *Origins* 23:12 (September 2, 1993): 207–211.

TAPES

"The Changing Church." *Thomas More Association* 1975.

"Apostolic Spirituality: A Challenge to Interiority and the Meaning of the Kingdom." *St. Anthony Messenger Press* 1976.

"Stewardship and Christological Ministry." *National Catholic Reporter Publishing Co.* 1988.

"We Also Show the Way." *National Catholic Reporter Publishing Co.* 1988.

"Contemplation: Finding Its Prophetic Voice." *National Catholic Reporter Publishing Co.* 1988.

Contributors

RONALD BARNES, SJ is Professor in the Pastoral Department at Regis College, Toronto and the rector of the Jesuit community.

GREGORY BAUM is Professor Emeritus at McGill University's Faculty of Religious Studies. He is the editor of *The Ecumenist*. His most recent book is *Karl Polanyi on Ethics and Economics*.

THOMAS BERRY, CP is the retired Director of the History of Religions Program at Fordham University and author of *The Dream of the Earth*.

ELAINE BIOLLO, SC recently received her Ph.D. in theology from Regis College, Toronto.

ANNICE CALLAHAN, RSCJ teaches systematic theology and spirituality at the University of San Diego. Her most recent book is *Evelyn Underhill: Spirituality for Daily Living*.

JOAN CHITTISTER, OSB is a member of the Benedictine Sisters of Erie, an internationally known author and lecturer, and executive director of BENETVISION: A Resource for Contemporary Spirituality. Her recent books include *Heart of Flesh* and *In Search of Belief*.

JOANN WOLSKI CONN is Professor of Religious Studies at Neumann College, Aston, Pennsylvania and editor of the newly revised edition of *Women's Spirituality*.

LEE CORMIE is Professor of Liberation Theologies at the University of St. Michael's College, Toronto and author of numerous articles in liberation theologies, church social teachings, and social movements.

CONSTANCE FITZGERALD, OCD is a Carmelite Nun in Baltimore, Maryland whose major work and study aim at the reinterpretation of John of the Cross and Teresa of Avila. She is author of "Impasse and Dark Night."

ROGER HAIGHT, SJ teaches systematic theology at Weston Jesuit School of Theology and is former president of the Catholic Theological Society of America.

PATRICIA COONEY HATHAWAY is Associate Professor of Theology and Spirituality at Sacred Heart Major Seminary in Detroit, Michigan.

MARY ANN HINSDALE, IHM is Associate Professor and Chair of the Religious Studies Department of the College of the Holy Cross. She is co-author of *"It Comes from the People": Community Development and Local Theology*.

SHARON HOLLAND, IHM is Office Head at the Congregation for Consecrated Life in Rome and Professor of Canon Law and Dean of the English Section at Regina Mundi Institute.

THERESA F. KOERNKE, IHM is Assistant Professor of Theology at Washington Theological Union, Washington, D.C., and adviser to the Bishops Committee on the Liturgy (NCCB) and the International Commission on English in the Liturgy.

MARY JO LEDDY is a journalist who lectures part-time at Regis College, Toronto. Her most recent book is *At the Border Called Hope: Where Refugees are Neighbours*.

ELLEN M. LEONARD, CSJ is Professor Emerita of Systematic Theology, University of St. Michael's College, Toronto and author of *George Tyrrell and the Catholic Tradition*, *Unresting Transformation: The Theology and Spirituality of Maude Petre* and *Creative Tension: The Spiritual Legacy of Friedrich von Hügel*.

MARY HEATHER MACKINNON, SSND is Director of the Doctor of Ministry Programme at the Toronto School of Theology. She is co-editor with Moni McIntyre of *Readings in Ecology and Feminist Theology*.

MARY MCDEVITT, IHM is Pastoral Consultant at Catholic Children's Aid Society, Toronto.

MONI MCINTYRE is Assistant Professor of Theology at Duquesne University, Pittsburgh, Pennsylvania and co-editor with Mary Heather MacKinnon of *Readings in Ecology and Feminist Theology*.

Amata Miller, IHM is an economist teaching at The College of St. Catherine in St. Paul, MN. She has taught and written on Catholic social teaching and economic justice issues for over two decades.

Susan Rakoczy, IHM teaches systematic theology and spirituality at St. Joseph's Theological Institute, Cedara, South Africa, where she is also engaged in the formation of spiritual directors. She edited *Common Journey, Different Paths: Spiritual Direction in Cross-Cultural Perspective.*

Sandra M. Schneiders, IHM is Professor of New Testament Studies and Christian Spirituality at the Jesuit School of Theology and the Graduate Theological Union in Berkeley, California and the author of *The Revelatory Text: Interpreting the New Testament as Sacred Scripture* and *Written That You May Believe: Encountering Jesus in the Fourth Gospel.*

Mary Ellen Sheehan, IHM is Professor of Theology at the University of St. Michael's College, Toronto. Recently, she completed a term as Director of the Doctor of Ministry Programme at the Toronto School of Theology. She has published several articles on issues related to women and ministry.

Carl Starkloff, SJ is Professor Emeritus at Regis College, Toronto School of Theology, and Associate Editor at the Institute of Jesuit Sources, St. Louis, Missouri. He is also an Instructor at Anishinabe Spiritual Centre in Espanola, Ontario.